NEW WOMAN ECOLOGIES

Under the Sign of Nature: Explorations in Ecocriticism

Serenella Iovino, Kate Rigby, and John Tallmadge, Editors

Michael P. Branch and SueEllen Campbell,
Senior Advisory Editors

NEW WOMAN ECOLOGIES

From Arts and Crafts
to the Great War and Beyond

ALICIA CARROLL

University of Virginia Press
CHARLOTTESVILLE AND LONDON

University of Virginia Press
© 2019 by the Rector and Visitors of the University of Virginia
All rights reserved

First published 2019

9 8 7 6 5 4 3 2 1

Library of Congress Cataloging-in-Publication Data
Names: Carroll, Alicia, 1960– author.
Title: New woman ecologies : from arts and crafts to the Great War and beyond / Alicia Carroll.
Description: Charlottesville : University of Virginia Press, 2019. | Series: Under the sign of nature | Includes bibliographical references and index. | Identifiers: LCCN 2018054024 (print) | LCCN 2018058047 (ebook) | ISBN 9780813942834 (ebook) | ISBN 9780813942810 (cloth) | ISBN 9780813942827 (pbk.)
Subjects: LCSH: Ecofeminism—Great Britain—History. | Ecofeminism in literature—History. | Women and the environment—Great Britain—History.
Classification: LCC HQ1194 (ebook) | LCC HQ 1194 .C37 2019 (print) | DDC 304.2082—dc23
LC record available at https://lccn.loc.gov/2018054024

Cover photo: Members of the Women's Land Army thistle hoeing. (Military History Collection/Alamy Stock Photo)

For Valera Carroll

Ideas survive those who give them birth.
—Helen Georgiana Nussey, *London Gardens of the Past*

CONTENTS

List of Illustrations ix
Acknowledgments xi
Chronology xiii

Introduction 1

1. Elemental Ecologies
   Arts and Crafts Women and Early Green Thought 27

2. "We Are Two Women"
   Alternative Agriculture, the New Woman, and Rural Modernity 60

3. The New Life and the New Woman
   Utopian Ecologies in New Woman Writing 84

4. "God Speed the Plough and the Woman Who Drives It"
   Ecologies of the Great War 118

5. Working Relationships
   Ecological Futurity and the Herbal Revival 152

Epilogue 189

Notes 195
Bibliography 221
Index 237

ILLUSTRATIONS

Figure 1. "Intensive Culture for Flat-Dwellers," *Punch*, 1917   8

Figure 2. *Oiketicus townsendi*, Olive J. Cockerell, *The Entomologist*, 1913   20

Figure 3. "The Seeds of Love," from Mary De Morgan's *On a Pincushion*, 1877   33

Figure 4. "The Pool and the Tree," Olive Cockerell, from Mary De Morgan's *The Windfairies*, 1900   45

Figure 5. "The Rain Maiden," Olive Cockerell, from Mary De Morgan's *The Windfairies*, 1900   52

Figure 6. "We are two women," from Olive Cockerell and Helen Nussey's *A French Garden in England*, 1909   64

Figure 7. Planting seedlings, from Olive Cockerell and Helen Nussey's *A French Garden in England*, 1909   72

Figure 8. Mending bell glasses with white lead, from Olive Cockerell and Helen Nussey's *A French Garden in England*, 1909   74

Figure 9. "Bacteria Thrown Out of Work," from Olive Cockerell and Helen Nussey's *A French Garden in England*, 1909   77

Figure 10. Women's Land Army recruitment poster, 1917   125

Figure 11. "Rosemary," from Lady Rosalind Northcote's *The Book of Herbs*, 1903   160

Figure 12. Herbstrewer's outfit   191

ACKNOWLEDGMENTS

This book was enabled by the richness of new approaches to ecocriticism and the depth of the subject. My biggest expression of gratitude must be given to those whose work made up the latter. As I came to know these impressive and complicated women, I realized I had to live up to them, and, since it was often shaped under duress, their work became more and more relevant to our own ecological crisis. While undertaking the project, I called upon and was answered by the goodwill of my family, students, administrators, archivists, colleagues, and most especially Auburn University's outstanding librarians, named below. Many of the latter even shared my excitement at locating yet another book, image, or article that built a case for New Woman ecologies.

Without the help of such people the book would not have been written. I'm grateful for my family's support. My husband Rob and daughter Alex have provided love and understanding each and every day. My siblings, Elizabeth and John Carroll, cheerfully acted as my personal art historians and consultants. My earliest mentors, Tom Lewis at Skidmore College and Fred Kaplan at the Graduate Center of the City University of New York (CUNY), have been strong supporters and friends. Friends and colleagues such as Claire Wilson, Jon Bolton, Paula Backscheider, Anton di Sclafani, Rajiv Mohabir, Sunny Stalter-Pace, Erich Nunn, Peter Logan, Penny Ingram, Chris Keirstead, and Marilyn Pemberton generously shared my excitement for this project. My students Caitlyn Anderson, Taylor Bowman, Brooke Bullman, Annie Gilbertson, Justin Paxson, Bryan Williams, and Robyn Miller have provided an ecocritical community for which I am extremely grateful. My chair at Auburn University, Jeremy Downes, assisted in locating travel funds, which made archival work and conferencing possible.

I'm grateful to my CUNY colleague Talia Schaffer for connecting me to Elissa Meyers, who helped with archival work at the Berg Collection in New York. Likewise, I'm grateful to Craig Bertolet for connecting me to Selina Dukes and Erika Roberts, who helped with archival work at the City of Westminster, in London. Auburn librarians Nancy Noe, Jaena Alabi, and Kara R. Van Abel at University of Alabama at Birmingham have been essential to this book and I am extremely grateful for their help.

Both the US and UK-I Associations for the Study of Literature and the Environment (ASLE) conferences have provided me with a forum for my work. It was at ASLE in the United States that I met Lowell Duckert, who very generously shared his introduction to *Elemental Ecocriticism* with me. Likewise, I'm very grateful to Dennis Denisoff for his editing of my article on Mary De Morgan for the *Victorian Review*. That became the seed of this project. Near the end, I received extremely helpful feedback from Samantha Walton, editor, with John Parham, of ASLE UKI's *Green Letters* on my article on Agatha Christie and medicinal plants, which is revised here as part of the final chapter. The British Women Writers Association, especially Roxanne Eberle, has also provided a community for my work and wonderful feedback.

At the University of Virginia Press, I am most grateful to acquisitions editor Boyd Zenner, who waited patiently for this manuscript and moved swiftly to bring it to press. I thank those who worked closely on the manuscript—Morgan Myers, Emily Shelton, Ellen Satrom, and the art department—for their help. I am also grateful to the editors of the series Under the Sign of Nature and my two readers, who provided insightful comments and excellent advice.

Finally, I'm especially grateful to my friend, gardening mentor, and mother, Valera Carroll. Her rescuing of local plants from construction zones was an early influence. That, and her love for art and textiles of all sorts, surely appears in these pages. The mistakes, on the other hand, are entirely my own.

Portions of chapter 1 were first published as "The Greening of Mary De Morgan: The Cultivating Woman and the Ecological Imaginary in "The Seeds of Love," in *Victorian Review* 36, no. 2, Natural Environments (Fall 2010): 104–17. Copyright © Victorian Studies of Western Canada. Reprinted with permission by Johns Hopkins University Press. Portions of chapter 5 were first published as "'Leaves and Berries': Agatha Christie and the Herbal Revival," in *Green Letters: Studies in Ecocriticism* 22, no. 1 (2018). © Association for the Study of Literature and the Environmental (ASLE UK-1). Reprinted by permission of Taylor & Francis, Ltd., on behalf of the Association for the Study of Literature and the Environmental (ASLE UK-I).

CHRONOLOGY

1879	Grain prices collapse, and the English agricultural depression begins
1891	Swanley Horticultural College admits women
1894	Sarah Grand publishes "The New Aspect of the Woman Question"
1898	Whiteway Colony established
1898	Lady Warwick Hostel for women's agricultural training opens
1902	Swanley Horticultural College closes to men
1903	Studley College founded by Lady Warwick
1903	Lady Northcote publishes *The Book of Herbs* in the Practical Gardening Series published by John Lane at the Bodley Head
1910	Lady Warwick presents *A French Garden in England* to King Edward VII
1913	Olive Hockin sentenced to four months imprisonment for suffrage-related arson attack on the Roehampton Golf Club
1914	The Pankhursts suspend suffrage activism and urge women to support the war effort
1914	The Whins Medicinal and Commercial Herb School and Farm founded by Maud Grieve
1916	Women's Land Army formed
1919	First herb garden exhibited at the Chelsea Flower Show
1931	Publication of Maud Grieve's *A Modern Herbal*
1939	End of the agricultural depression
1941	Passage of the Pharmacy and Medicines Bill

NEW WOMAN ECOLOGIES

INTRODUCTION

In November 1906, Octavia Hill (1838–1912) took a break from the greening of London to write to her goddaughter, Olive Juliet Cockerell (1869–1910). Hill was deeply worried about Cockerell, an illustrator trained in the Arts and Crafts school. In what seemed to be a "complete reversal of occupation," the young woman was moving to the country.[1] There she would train to become a market gardener, eventually growing and selling fresh local fruit and vegetables in partnership with another woman, former hospital almoner Helen Nussey (1875–1965). Cockerell and Nussey meant to see if "two women like ourselves" could "make a living in a 'small holding,'" practicing the French tradition of intensive culture with "no labor but our own."[2] The illustrated book they left behind, *A French Garden in England: A Record of the Successes and Failures of a First Year of Intensive Culture* (1909), suggests they succeeded. Nonetheless, Octavia Hill, the champion of the urban open space movement, was nearly inconsolable over her goddaughter's shift to "country work," pleading with her to "pray beware of potato digging & over fatigue!"[3] While Hill could well understand and encourage her niece to participate in the lifting of dirt from the London slums, she could not support the lifting of soil by a lady for the purpose of growing and selling food. Nor could she understand what she saw as the mystifying, self-imposed exile of two such vital young women from London. They would be "sorely missed," she said, in the metropolis.[4] For Cockerell, however, her garden future seemed bright. Living out the elemental ideals of the Arts and Crafts movement from the ground up, her hands would cultivate the earth itself. She would fulfill the now often forgotten principle of early green socialism: "England should feed her own people."[5]

This book reclaims the intersection of such initiatives with the literature and culture of the New Woman. From the Arts and Crafts period to before, during, and after the Great War, the discursive image of the latter informed historical women's responses to the keen environmental debates and crises of their day. These include familiar concerns about air and water quality, as well as now less familiar critiques of Victorian floral ecologies, extinction narratives, land use, local food shortages, and food importation. Indeed, as importation squeezed traditional British farm crops and labor practices under laissez-faire economics, the great agricultural depression began, arcing over the period of the

New Woman, from 1879 to 1939. This crisis created opportunities for women seeking to remake both the land and themselves for the future. It accelerated the third wave of "alternative agriculture" in British history manifested in Cockerell and Nussey's small, local market garden.[6] The Land Question, a hotly debated social interrogative contemporary with the Woman Question, emerged during this period as more conventional monoculture farms went bankrupt. The crisis concerned early greens and progressives, as well as conservatives. The former, in particular, challenged both the economic losses to farmers and the "appalling spectre of the staple trade of humanity, the production of food, in a state of threatened dissolution in England."[7] As the government had "taken no action except to gaze upon the disquieting spectre" and bear "the spectacle with apparent equanimity," many argued the benefits of going back to the land themselves.[8] For women, speaking from the land became a way, as militant feminist turned war-time landworker Olive Hockin wrote, of proving women could do "any mortal thing" they wished with their minds and bodies.[9] Such a claim extended New Woman polemicist Mona Caird's 1888 demand for the "acknowledgement of the obvious right of the woman to *possess herself body and soul*" in marriage to an urgent new context.[10] Testing this New Woman discourse on autonomy, historical women like Cockerell, Nussey, Hockin, and many more made, or represented, material contact with a vibrant, more-than-human world. Both marking and marked by the latter, they experienced and imagined immanence entirely apart from heteronormative reproduction.

Now rarely discussed in the same breath, the Land and the Woman Questions converged in a variety of new contact zones in which women conceptualized an early green modernity. Women involved in or influenced by the culture of the Arts and Crafts movement, for example, contrasted earlier land-use practices to those of their own time by reading early modern texts such as Gerard's *Herball or General Historie of Plants* (1597). Led by the example of William Morris, they sought to regain intimacy with the land through their deliberately archaic use of plants and the elements in their themes, materials, and practices.[11] Others, among the first middle-class women to go on the land by entering agricultural programs, utopian "New Life" agricultural communities, or the Women's Land Army, saw that the current economic crisis was also what we would now term an ecological one. They understood that, as farmland went fallow and weeds encroached on large farms designed to support monoculture, local places lost a full spectrum of edible and medicinal plants as well as the once thriving culture of plant literacy that supported them. Women herbalists after the Great War saw "no reason" why biodiverse culture should not be revived to profit women and the land in England.[12] Many took advantage of Olive Schreiner's mandate to the New Woman: "Take all labor for our province!"—including "the non-sexual fields of intellectual or physical toil.[13] The result was the formation of ecologies, situated systems of knowledge born out of women's often self-interested, vocational interventions into what we now call the Anthropocene. These naturalized both an ecological modernity and the New Woman's privilege to access the same. A way had been paved, in fact, by New Woman literature and visual culture that interrogated and reinvented a deeply entrenched, gendered topos of Nature inherited from earlier generations. The majority of the women studied in this book challenged the latter with fresh alternatives, fostering now recognizably ecological models avant la lettre. As they greened the sexual anarchy of the New Woman, they gendered early environmentalism.

Inevitably, studies of the New Woman lead to questions about the term itself. "The New Woman," writes Regenia Gagnier, "was the term applied to self-consciously modern women at the fin de siècle."[14] In their case, modernity meant claiming "the pursuit of material well-being and economic independence, scientific knowledge, and political emancipation."[15] Such claims, Elaine Showalter notes, historically allied the New Woman with a threatening sexual anarchy as she sought "new opportunities for education, work, and mobility" as well as "alternatives to marriage."[16] Talia Schaffer specifies that "this dangerous 'New Woman' was a middle-class woman agitating for such 'dynamite' ideas as the right to walk without a chaperone, to hold a job, to live alone in a flat, to go to college, and to wear sensible clothing."[17] But each of these critics counters that the term "New Woman" was "always contested," not least by women themselves.[18] As early as 1999 and as recently as 2013, Sally Ledger and Talia Schaffer have noted that contemporary critics continue to find the question of how to define the New Woman "vexed."[19] Ledger notes that even as the New Woman was named pejoratively in Victorian print culture, her image opened a "discursive space" that "was quickly filled by feminist textual productions sympathetic" toward demands made by Victorian feminists.[20] Even so, from Mona Caird to Sarah Grand, historical women's definitions of the New Woman and the futures they envisioned for her, as well as the ideological standpoints from which such futures were imagined, vary widely. There must be "some core values" through "which we can identify" the New Woman, Schaffer writes, and perhaps as well "a particular way of writing " that characterizes her.[21]

My hope is that both those core values and way of writing are represented here. My subjects share what Gagnier describes as the "essential" feature of New Woman literature—the search for autonomy: "What New Women wanted, collectively, was freedom, autonomy, not 'power over,' but 'power to,' empowerment."[22] I argue that this gendered project intersects with early green socialism and historical events such as the agricultural depression and the formation of the Women's Land Army during the Great War. Gagnier contends that "what clearly emerges in New Woman literature is the difference between independence or separateness and autonomous individuals in relation."[23] In their book, then, Olive Cockerell and Helen Nussey negotiate their freedom as "two women." They manage their distance from, and relation to, the city and the local village, their home and work, their partnership and their concerned families. However, in their garden, other relations and actants besides human ones also challenge their ideal of autonomy. My study extends the New Woman's engagement of the latter to her quite specific historical and material encounters with an agentic, more-than-human world. The writing, material culture, gardens, and art that result share, if not one way of writing, painting, gardening, or crafting, then one self-consciously modern, frankly experimental ethos. From *A French Garden in England* (1909) to Maud Grieve's *A Modern Herbal* (1931), the resulting work explores how new professions for women shaped ecologies in the shadow of war, economic crisis, and environmental degradation. It also records how women, seeking to prove their autonomy from the ground up, uncovered powerful ecological systems and actants that challenged human exceptionalism itself.

Studying the New Woman's participation in the ecological story of her time reveals a new perspective on familiar subjects like Sarah Grand (1854–1943) and her infrequently read novel on intensive culture, *Adnam's Orchard* (1912), or on George Egerton's tale of a rural women's utopian craft community, "The Regeneration of Two." It also introduces a host of new subjects often unnoticed by New Woman scholars and ecocritics. These women have rarely been considered English nature writers, nor are they included in anthologies of the same. But they are nonetheless a rich source of ecological thought. They include utilitarian gardeners like Maud Grieve (1858–1941) and Eleanour Sinclair Rohde (1881–1950); Arts and Crafts writer Mary De Morgan (1850–1907); Great War–era landworker and poet, Rose Macaulay (1881–1958); crime-fiction writer Agatha Christie (1890–1976); founder of Studley College, Frances Evelyn Greville, Countess of Warwick (1861–1938); Newnham College historian of botany Agnes Arber (1879–1960); artists like Cockerell, May Morris (1862–1938), and Olive Hockin (1881–1936); and "New Life" communitarians Nellie Shaw (1877–1939) and Edith Ellis (1861–1916), who wrote under her married name, Mrs. Havelock Ellis. They are joined by named and anonymous members of the Women's Land Army of the Great War who left records of their encounters with the land in their own journal, the *Landswoman*.

Their contribution is timely, addressing issues such as food justice and importation, equitable land use, environmental toxins, biodiversity, and the agency of the material world itself. To read their work, I draw on recent insights from feminist new materialisms, as well as from queer and elemental approaches to ecocriticism. The first of these elucidates the central irony of this project: that as historical women test the New Woman discourse of "absolute bodily autonomy"—by going on the land, for example—they position themselves to encounter a wide range of vibrant, more-than-human bodies.[24] Stacy Alaimo's theory of trans-corporeality, Jane Bennett's *Vibrant Matter,* and Cohen and Duckert's *Elemental Ecocriticism* act as lenses through which to read plants, soil, the elements, and women's sometimes exhilarating, sometimes disturbing new knowledge of the same. Such perspectives illuminate unexpected alliances and even unwelcome traffic between bodies as middle-class women record or imagine their historic "female firsts" on the land, in the elements, and amid more-than-human nature.[25] Valuable and underappreciated as texts such as Olive Hockin's memoir *Two Girls on the Land: War-Time on a Dartmoor Farm* (1918) may be, they benefit from readings that apply pressure to the "triumphant" readings of the New Woman in second-wave feminism.[26] Reading materially allows a close analysis of the ways in which women's discourses, even of autonomy, may be implicated in an unsustainable human exceptionalism and class privilege. New materialist critiques of the human value of transcendence over immanence inform my readings of women's explorations of hierarchies that, once thought vertical, are found to be horizontal. As the linguistic and constructionist turn of feminist theory produced powerful readings of the New Woman as a discursive figure, so does a materialist reading of the intersections of discourse and matter produce important insights into New Woman literature. Reading the wind or rain in Mary De Morgan's tales or Rose Macaulay's poetry reveals that such elements challenge the very idea of boundaries between bodies and between nations. Reading plant life in women's herbal texts of the early twentieth century makes quite clear that what has been assumed in modernity to be "inert" retains its potency.[27]

This is certainly the case in Cockerell and Nussey's small-space, raised-bed garden, the vitality of which defies its small acreage as well as gendered and national boundaries. The material dimensions of the garden itself with its practice of dense companion planting reveal what Nancy Tuana calls "the rich interactions between things" rather than the meetings of stable essences.[28] In such a garden, it becomes clear that "subjects are constituted out of relationality" and that their process of becoming may occur through engagements of bodies of all sorts.[29] This book will explore how middle-class women form rural communities and single-sex partnerships that mesh with the "exuberant pleasures of thinking with, and feeling with, an abundantly, uncontainably queer world."[30] Women's observations of plants that reproduce aided by other species, for example, both challenge widely accepted competitive Darwinian narratives of the struggle for existence and queer the ontological boundaries placed between living things. Freshly exposed to this exuberance in biodiversity, women's love for other women seems suddenly . . . natural. As the science of ecology develops alongside new readings of Darwin, it leads to foundational critiques of the latter in women's literature and practices. These intersect with early green interests in, for example, the idea of mutualism over competition as the basis of evolutionary life.[31] In their engagement with the discussion of mutualism and equality for women, New Life followers Nellie Shaw and Edith Ellis, for example, were compelled to conduct their own experiments in collective or cooperative living. Involved, respectively, with the communities of the Fellowship of the New Life and the anarcho-communist Whiteway Colony, both pursued rural social experiments that found collective agriculture "the basis of all constructive work."[32] They sought to construct the future in the present. Work that came out of such cooperative social experiments, such as Ellis's neopastoral *The Lover's Calendar*, may deploy utopian practices to disrupt what José Esteban Muñoz theorizes as "straight time," seeking to replace it with a queer ecological futurity.[33] Following such perspectives, this book works with new ecocritical approaches to "rematerialize the social and take seriously the agency of the natural," reading both the discursive work of the image of the New Woman and the material encounters of historical women. The "dynamic and interactive nature" of both, "diffusely enacted in complex networks of relations," is of course clear in all literature, but it is particularly at work here, given the convergence of the New Woman discourse on autonomy with the ecological issues and historical events of the time.[34]

The intersection of New Woman culture with the latter, I stress, can produce disturbing results. It is clear that most of the women studied here object to the outsourcing of food based on personal or national self-interest rather than in concern for the impact of food production on other lands and peoples. Agricultural historian Nicola Verdon cites Edith Bradley of Lady Warwick's historic agricultural training program for women as evidence of the latter. She "urged, 'get the women on the land to rear the necessary poultry and eggs, and so let them have the benefit of at least some of the money now enjoyed by the foreigner.'"[35] Moreover, the role of fitness endemic to the discussion of mobility and bodily autonomy in New Woman literature and culture often finds its source in the pseudo-science of eugenics, an ideology popular with writers

Edith Ellis, George Egerton, and Sarah Grand, as well as suffragette Olive Hockin. As women claimed the land for the people, they often exerted their privilege over others, leading us to ask, Which land was intended for which people, and who decides? Likely influential in the earliest forms of environmentalism, preservation and conservation in England, women exerted their ideas of heritage to varying degrees, seeking to shape the role of the land in the future as they saw fit. These aspirations are clearly stated through what Gail Cunningham identifies as "New Woman principles of plain speaking and rigorous self-analysis" and deployed through visual and print culture to make women's voices heard.[36]

Indeed, even as the icon of the mainly urban New Woman emerged in print culture during this period, it constructed and reflected mainly middle-class women's rethinking of an abstract Nature and their potential new privilege to enter traditionally masculine spaces. The latter include laboratories, studios, commercial and collective gardens, print culture, and literal fields, such as the one in which the horrified Octavia Hill imagined her goddaughter digging potatoes. Such interests became fodder for *Punch*, appearing in cartoons such as "Intensive Culture for Flat Dwellers," where a New Woman figure in breeches and bobbed hair spreads "early mustard and cress" seeds on a doffed petticoat in her parlor (see fig. 1). Glass cloches cover the tables and a pig peeps out of a china cabinet. A bemused gentleman lurks slightly off to the side, meekly watering what the woman plants. This type of cartoon clearly expresses the anxiety elite men felt about New Woman ecologies, their disruption of formerly stable gendered binaries like nature and culture, city and country, inside and outside, production and reproduction, working and ruling classes. These eroded the solid ground of the patriarchy; in the cartoon we witness the material transformation of human underwear into food-growing soil. Clearly, the New Woman grower presents the arrival of a disturbing green modernity.

While such images were meant to mock the idea of women's entry into male spaces or professions, historical women "acknowledged and even embraced the body *Punch* concocted for [them] as an epitome of liberation. By its relentless caricature of woman as athlete," for example, "the characteristics that would later be recognized as 'New Woman qualities,' coalesced."[37] "Intensive Culture for Flat Dwellers" similarly represents an appealing, highly mobile, transformative Nature and an elegant urban woman's intimacy with the same, quite apart from reproduction. The doffed petticoat is serving a better purpose, replaced by a stylish and comfortable outfit. This image of the gardening woman is in fact not far off from the illustrations of *A French Garden in England*, with one important difference: those authors exclude men nearly

Figure 1. "Intensive Culture for Flat-Dwellers. Sowing Early Mustard and Cress on Winter Underclothing." *Punch, or the London Charivari,* vol. 152, 30 May 1917. (Courtesy of the Mervyn H. Sterne Library, University of Alabama at Birmingham)

entirely. Their illustrations carefully highlight their partnership, showing them digging, plunging their hands into the soil, planting seeds, and handling substantial cloches and glass windows over raised beds. They represent themselves as perfectly natural, always in motion, laboring together in jaunty ties and hats. The figure of the New Woman, although existing mainly in the "foolscap and ink" of print culture, was potentially useful to such historical women seeking to enter masculine space and to those who desired other women to do the same.[38] Lady Warwick, for example, the patron of the women's horticultural and agricultural program that Cockerell and Nussey attended, wrote in glowing terms of joining "the New Women and Old Acres" in the college's journal, the *Women's Agricultural Times*.[39] The Women's Land Army during the Great War used the image and vocational discourse of the New Woman as a recruiting tool, as did advocates of women's herbalism during and after the war. As historical women engaged the keen environmental crises of their day, they constructed ecologies that intersected with the figure of the New Woman, a paradoxically nonessentialized figure influencing systems of knowledge they developed and the gendered opportunities women created on land perceived as widely "neglected" during the agricultural depression.[40]

Such women may have been smaller in number than the teachers, welfare workers, and secretaries studied in groundbreaking studies such as Martha Vicinus's *Independent Women: Work and Community for Single Women, 1850–1920* or Gillian Sutherland's *In Search of the New Woman;* however, the literature and culture they produced is substantial and timely, including an interrogation of practices like floral gardening, large-scale industrialized farming, and monoculture. Examples of women's agricultural partnerships are only now being brought to light and interesting perspectives on statistics are emerging. The number of women farmers in England, for example, held steady from 24,338 in 1851 to 21,548 in 1901, despite the overall decline in farmers due to the agricultural depression.[41] As growing and farming "came to be seen as a viable occupation for middle-class women in England" before the Great War, "never married or single women formed 27 and 29 percent of all women farmers" by 1931.[42] This suggests that as women sought new opportunities in horticultural or agricultural training during what is traditionally considered the age of the New Woman (roughly 1870 to 1914), they reaped longterm benefits. Often, learning forms of "alternative agriculture" such as intensive culture seems to have served women well, even as traditional monocultural crops continued to struggle to get a good price.

Indeed, although now recognized by ecocritics and activists like Vandana Shiva as anathema to biodiversity and as the chief polluter of all industries, as early as 1914, industrialized farming and monoculture struck some women—such as poet-turned-landworker Rose Macaulay—as a counter to biodiversity that came with a high social and ecological cost. Initially, women's rethinking of such institutions was often inspired by their schooling or the early green simplification practices of the late nineteenth and early twentieth centuries stemming, as in Cockerell's case, from "green Victorians" like John Ruskin and William Morris.[43] To this influence, however, they added their own, often scientific, interest in soil, intensive culture, and herbal medicine. Open to technology such as the raised bed and glass, their work may be understood through Shiva's biodiversity paradigm as "a biodiversity-based productivity framework" rejecting the failing triad of beef, wheat, and dairy and replacing it with a growing strategy that "reflect[s] the health of nature's economy and people's economy."[44] Their work, therefore, is relevant even now, displacing both destructive agricultural practices and what Shiva terms a mental monoculture blind to "the ecological functions arising from the relationships and cooperation between diverse living components of an agroecosystem."[45] These come vividly to life in texts like Maud Grieve's *A Modern Herbal* (1931), a still-definitive, monumental text that envisions many hundreds of plants as potent partners to human health and wellness in modernity. As the nation was cut off from importation during the Great War, women herbalists were to rediscover the benefits of this biodiversity by necessity. They learned viscerally and materially where their breakfast and medicine came from. As Timothy Morton acknowledges in *The Poetics of Spice*, "To think about this . . . is to uncover global networks of power."[46] During the Great War and the agricultural depression, tracking local food or plants and their presence, absence, wellness, or neglect was tracking the environmental injustice of empire, ironically often perceived as most destructive at home. Reading the experiences of women's immanence through their encounters with plants, cold, heat, and animals allows the local and global, "material and discursive, natural and cultural, biological and textual," to be seen as "entangled territories."[47]

Shaped within these new contexts and companions, New Woman ecologies as a subject provides an opportunity to move forward to a feminist ecocriticism that deconstructs the gendered material/discursive binary, ideally retaining "both elements without privileging either."[48] This book unpacks the gendered and classed discourse that shapes Octavia Hill's shocked reaction to her niece "digging potatoes!" and details the impact of that material encounter on multiple bodies. Developing such intimacies between women and matter forms new perspectives on the earth. It may suddenly be perceived as an actant, an ally, or as a force capable of "unpredictable and unwanted actions" delivered by the elements, weeds, chemicals, or systems themselves.[49] To see it as the latter is to release an abstract Nature from its role as nurturing or vengeful mother and to place it in a far more potent role by studying its materiality itself. If discursive formations of Nature have long been "waged" as a "cultural repository of norms and moralism" against "women, people of color, indigenous peoples, queers, and the lower classes," a new materialism may locate matter as something less compliant, a "vibrant" thing in and of itself.[50] This ecological legacy, rarely acknowledged, becomes part of the material and discursive work of New Woman literature and culture, ironically enabled by women's interrogation of an abstract Nature itself.

"Whatever Is Natural Is Certain to Be Wrong": Nature and the New Woman

In New Woman texts such as "Phases of Human Development" (1894), polemicists like Mona Caird (1854–1932) state with great conviction that "whatever is natural is certain to be wrong."[51] "Every progressive being has" to "evade"

"Nature," Caird explains.[52] In her analysis of a gendered, discursive Nature, Caird argues that ideas of the same are deployed to constrain women to the role of reproduction. She sees what ecofeminist Val Plumwood and so many other contemporary ecocritics argue: that Western culture has constructed a female Nature in opposition to a masculinist culture. Ecological feminists such as Plumwood note that this "master model" resists "the recognition of dependence," ordering itself conceptually in terms of a male (and truly human) sphere of free activity taking place against a female (and natural) background of necessity."[53] This move elevates reason and masculinizes the trait, creating a feminized Nature and constructing a "culture of mastery" that transcends "her."[54] As women are linked to the physical, the reproductive, and the earth rather than to the mind, the productive, and the cultural, they are made immanent, grounded in matter. Once there, they are debased by a cascade of interlocking gendered, raced, and classed dualisms that support the nature/culture divide. The purported "naturalness" of the reproductive woman, the intellectual inferiority and emotionalism of women, and their inferior physical strength are all linked conceptually to this hierarchical culture/nature discourse.[55] Hence, in the Victorian period, widely published pseudoscientists like Herbert Spencer argued that women's bodies are best limited to reproduction because their mental faculties are inferior to those of men.[56]

That burden of reproductive immanence creates a dilemma in New Woman fiction such as George Egerton's short stories, which often represent the more-than-human world as a sign of a beautiful, pleasurable, but inescapable reproductivity, a seductive trap that ensnares women, especially when they seek to explore sexual pleasure and even as they seek to enjoy their physical competence out-of-doors as sportswomen, travelers, and more. To uplift women, Caird fully embraces the denigration of the abstraction of the construct of Nature she knew. "By becoming less and less 'natural,'" she argues, "the human being becomes more and more tolerable."[57] By "diverging from the tame and barren wastes of 'Nature'" a new step of progress is taken.[58] Placing "Nature" in scare quotes, Caird's critique makes her audience see that this "Nature" is indeed merely a discourse. As such, it is "tame," subject to masculinist authority, and a "barren waste" devoid of material for constructing a New Woman. Disavowing the fetish is an imperative, an article of faith in the development and formation of a New Woman. Caird brackets "Nature" as illusory, arguing as Marx would that such abstractions conceal social relations linked to power, having "nothing whatever to do with the physical properties" of the thing itself or with "the material relations" arising from them.[59]

However, in this very disavowal of the fetish "Nature," Caird opens up radical possibilities for rethinking it and calling it by other names, such as those "material relations" and "physical properties" obscured by the fetish. These are very much of interest in New Woman literature and culture. If "Nature" was a fetish or illusion, the land, plants, elements, soil, and agricultural systems were not; they might be subjected to rigorous analysis and reenvisioning by women. Materially, they were vibrant potential allies, capable of naturalizing other subjectivities, such as the "invert" sexuality identified as "the best the world c[oul]d have for me" by Edith Ellis.[60] An interest in material relations and physical properties might guide an interest in land use, while land might, as in Cockerell and Nussey's garden, provide space for two women to work and love. The latter possibilities, moreover, might lead to entirely new perceptions of what Nature is. By questioning material relations and shifting physical properties quite literally, implementing a newly fit and competent woman's body or technology like glass and steam in Cockerell and Nussey's garden or rethinking the leisured class's abandonment of local plants in Sarah Grand's novel *Adnam's Orchard*, the merely reproductive fetish Nature—the notorious "botanical vernacular" of the culture of "bloom"—might be displaced.[61]

The material relations of places once associated with an abstract gendered Nature come under fierce review in Grand's novel and in her classic essays, like "The New Aspect of the Woman Question" (1894). There, Grand depends upon traditional monocultural agricultural imagery to epitomize the current subjection of women in England and to contrast them with men, who are fully human and enfranchised. Grand provocatively characterizes the majority of British men as "satisfied with the cow-kind of woman as being most convenient; it is the threat of any strike among his domestic cattle for more consideration that irritates him into loud and angry protests."[62] She argues that women as well as men are to blame for this social arrangement, the sex-gender system in which women are mere breeding stock subordinated to protective male figures. Relations on the traditional farm, which reify this false narrative, are critiqued and shifted in Grand's novel *Adnam's Orchard*, as she shapes a new kind of farm and garden and a new gendered ecology, embracing the scientific method of intensive culture Cockerell and Nussey deploy in their French garden. Her new all-green garden critiques not just the floral discourse of the gendered culture of bloom that Amy King analyzes, but its ecology and the Nature it supports.[63] This critique appears in the work of Mary De Morgan and the modern herbalists as well. They follow Caird's recommendation in "A Defense of the 'Wild Women,'" to be something other than a "'modest violet, blooming unseen,' unquestioning, uncomplaining, a patient producer of children and consumer of flowers."[64]

As New Women literally followed Olive Schreiner's example and recommendation in *Women and Labor* (1911) that educated, middle-class women "take all labor for our province!" including "the non-sexual fields of intellectual or physical toil,"[65] they engaged the idea that they too might go on the land. In texts like Schreiner's "Dream of Wild Bees" (1890), "the ethos that Schreiner envisions emphasizes a unity among all beings in order to promote equality."[66] Her *Story of an African Farm* (1883) locates connections between human and more-than-human nature as her heroine traces the bodies of animals and plants on the farm, noting the similarities across species. This moment of horizontal recognition is shaped, however, in colonized surroundings, on a remote location that stands in for wilderness. What would such intimacies between human and more-than-human nature look like in a toxic environment or in a postpastoral England where farms and local produce were in steep decline and where outsourcing of labor to places like South Africa had impacted local ecologies and food systems? How would English women engage this Nature so intimately if the production of the "staple of human life," food, had been lost at home?[67] In engaging these questions, British women sometimes assumed Schreiner's position—that is, that the land was theirs to interpret and occupy even if they were relatively recent arrivals.

In mainstream Victorian texts such as Grant Allen's anti–New Woman novel, *The Typewriter Girl*, these choices seemed a sign of degeneration, while the landscape itself seemed marked by decline. However, in illustration, art, gardens, and even in the production of new plants such as Seal lavender, a hybrid constructed by the partnership formed between two woman-run herb farms, material female partnerships note a still resilient world far more exuberant than previously understood through the marriage metaphors of Linnaean taxonomy. Ironically, while the city was purportedly the haunt of the New Woman and queer people, the country offered Cockerell and Nussey or Edith Ellis the opportunity to live "freely," sometimes with other women, "quite hidden in the woods," yet close to London.[68] Their work represents same-sex households that challenge traditional Natures and the heteronormativity of the pastoral. In the cases of Olive Cockerell and Helen Nussey, May Morris and Landswoman Mary Lobb, and Olive Hockin and her female wartime farm companion she renames "Jimmy," their households construct a queer system of knowledge that challenges gender and food injustice as well as the traditionally gendered farm labor in England.[69] Their experience with the latter, as well as with early green politics and New Woman literature and culture, had primed them to rethink Nature and the intersection of unfair labor practices with ecological destruction.

Arts and Crafts, Early Greens, and the New Woman

Indeed, New Woman ecologies intersect with the early green labor practices of women in the Arts and Crafts movement. From Mary De Morgan to Maud Grieve, such subjects often respect the "Ruskinian ideal of manual labor as a form of artistic expression and a manifestation of the worker's innate dignity."[70] They embraced the idea that seeking haptic contact with the elemental vibrancy of the natural world was antidotal to the work of industrialization, which exploited both people and the material world. In the tradition of early green politics, they also sought to simplify life so as to reduce waste and consumption collectively. De Morgan's anti-industrial tales, then, show the influence of William Morris. Now celebrated as "one of the most potent nineteenth-century expressions of environmental politics, ethics and aesthetics," Morris's public lectures and creative works argued for cleaner energy "such as timber, water and wind power" instead of coal.[71] In writing against the opening of a Kent coalfield—a "new manufacturing hell"—he recommended plans to increase food production. Such plans would deserve popular support because food was real wealth, to be used by all.[72] Morris's *News from Nowhere* offers hope for a future, agrarian society in which people achieve "a nature bettered and not worsened by contact with mankind."[73] For women like Lady Warwick, Cockerell and Nussey's mentor and the founder of Studley College, Morris was a key influence: "the perfect expression of the type of personality for which the Utopian impulse of the reformer is forever seeking."[74] Increasingly, Lady Warwick stressed Morris's demand that local food should be counted as wealth. During the Great War, she charged that the absence of a domestic harvest was not only immoral, but it made an ecologically damaged nation more warlike: "At present time," she writes in her memoir *A Woman and the War*, we "only produce about twenty-percent of the food we eat. For the rest, we depend upon our mercantile marine and our power to hold, not only the seas, but the skies above."[75] Warwick links the nation's faltering agrarian tradition to local industrialization as well as to global industrialized capitalism and food importation, pointing out the intertwining of the nation's ecological vulnerability and military aggression. With other friends of Morris, like the anarcho-communist Peter Kropotkin, she stressed the value of a vegetable diet and the benefits of intensive culture and rethought the high ecological cost of bread and the plough.

Although often associated with the preservation of historic buildings and the use of natural materials and themes, iconic Arts and Crafts leaders like Morris in *News from Nowhere* and Walter Crane in his design "A Garland for May Day" (1895), which appeared in the *Clarion*, reveal the importance of food justice to early green politics. This message unfurls from the banners sprouting from Crane's Flora figure, articulating key issues in the Land Question debate: "The land for the people," "No people can be free while dependent for their bread," "England should feed her own people," "The plough is a better backbone than the factory."[76] Each of these imperatives is foundational, placed at the base of the Flora figure and growing from the grass itself. Inspired by the great crisis of the agricultural depression,[77] the debates around the Land Question, and an early green "mission of restoring the People to the Land" at the turn of the century, some feminists, craftsmen, artists, and thinkers sought "the right to grow their own food and live simply."[78] At the same time, the depression itself provided new opportunities in what Joan Thirsk terms "alternative agriculture," the growing of herbs, vegetables, and fruit in market gardens as an alternative to unprofitable and unsustainable mainstream crops.[79] This work often appealed to those influenced by the Arts and Crafts movement who wanted not just to represent its ideals in art, but to live them.

Paradoxically, then, the emigration of farmers opened a niche for English women on the land. Why should they remain "in the dreary seclusion of" a "tiny city tenement" when "there is a green world beyond the gates, with woods and streams, from which we are shut out for want of opportunity?"[80] Scarcity of local food spurred women's existing sympathy with early green ideologies and led to the creation of new programs that taught intensive growing methods augmented by scientific knowledge and glass. This method of going "back to the land" also appealed to social radicals and welfare workers like Adela Pankhurst or Helen Nussey, as well as to Arts and Crafts–trained artists like Olive Cockerell or Olive Hockin.[81] Even before the war, such gendered endeavors were "news," traceable in newspapers and books published by major New Woman publishers such as W. T. Stead, who published Cockerell and Nussey's garden book.[82] In the latter case, the famed New Woman publisher joined forces with the book's Arts and Crafts printer, W. H. Smith & Sons, located in the experimental community of Garden City, Letchworth. The sensibility of the movement is thus expressed in the very binding and printing of the book as well as in its content.

In their books and in early modernist journals like the *New Age* and the *New Order*, early greens express grave concern about England's diminishing food stores and its loss of local plants in the early twentieth century. This absence made a niche for market gardeners, and women clamored for entry. Both Studley and Swanley Colleges and smaller programs sought to recruit unmarried, educated, already professional middle-class or elite women from urban locations to their agricultural and horticultural programs. There, women would be intellectual as well as agricultural leaders. Studley College might have the atmosphere of "Girton or Newnham" and link agricultural training for women to "economic policy."[83] Like other intentional and experimental early green programs and communities, such programs embraced "the theory that everyone should do his fair share of manual labor, that the workers should be thinking and that the thinkers should be working."[84] Food production could and should replace factory work for both sexes, and of course it could provide work for unmarried middle-class women. "Opportunities in horticulture posed solutions to the" agricultural depression and to the "so-called redundancy of single women, whose educational and occupational needs were championed by feminist leaders."[85]

Such schemes evoked an agricultural future, the New Life, pursued by Edward Carpenter and detailed in Peter Kropotkin's *The Conquest of Bread* (1906).[86] Influential socialist feminists of means like Lady Warwick seized the moment and shifted the potentially retrograde discourse of women and nature in partnership. She removed the image of the procreative woman from the scene entirely to associate "The New Women and Old Acres."[87] The new agricultural woman was intelligent, economically independent, fit, productive, technological, and scientific. At Studley and Swanley Colleges, women studied the science of soil composition, extended the growing season, and reclassed the laboring woman's body as intelligent, middle-class, and professional. Such women "would not flinch from an experiment."[88] Once women were limited to "pluck[ing] roses." Now Lady Warwick "marvel[ed] at the progress feminism has wrought in the world." Women might also dig a trench and "earth-up" plants. "Physical exercise could not" any longer "be limited to one sex."[89] Her work is energized by her refusal to accept women's limitations and the moral urgency of a burgeoning food-security crisis.[90] She too is dismissive of women's floral gardening, citing rose care as neither riveting nor useful.

While Victorian protoenvironmentalists like Octavia Hill had defined the greening of England as a project concentrated on developing green space and beautifying the spaces where working people lived, for Lady Warwick such initiatives did not go far enough. Hill did not seek to transform how people worked. Moreover, her vision of women's supervisory roles kept their hands clean. In 1899, when the International Congress of Women met in the city of Westminster to discuss the entry of women into the professions, it challenged that limit. The congress held three sessions on women and agriculture, and from these sessions twenty-two women emerged to form the Women's Agricultural and Horticultural International Union (WAHI) with the help of Frances Wilkinson (1855–1951) who was in 1904 to become the first female principal of Swanley Horticultural College in Kent. Lady Warwick then held her own meeting, forming a rival group with the support of Dr. Elizabeth Garrett Anderson, who testified to the fitness of middle-class women to work the land. The group declared a resolution "in favor of duly qualified women being afforded the advantages of full fellowship in scientific and other learned societies."[91] Profoundly challenging the division between city and country, such women deployed the resources of London to intervene in the countryside, which was then deep in the grip of the agricultural depression. They sought not to "return" the countryside to a former idyll, however, but to make it a proving ground for the New Woman.

The urban culture of the latter ironically aided in this project, providing convivial women's clubs as meeting places. From there, women put "their education, their professionalism, in a word their culture, at the forefront of their campaign" to put women on the land.[92] It was from London again that women were to organize the Land Army during the Great War. In fact, it was in London and other English cities where recruiters sought already independent and mobile urban women who might be transformed into landswomen. This intellectual and social link between city and country establishes the character of New Woman ecologies, which shrinks rather than widens the distance between city and country, deploying the idea of the New Woman's mobility in the city to construct her presence on the land. The famed herbalist Maud Grieve, for example, took full advantage of the research collection and plants in the British Library, Kew Gardens, and the Chelsea Physic Garden, and showed off her knowledge by entering the first herb garden exhibit in London's Chelsea Flower Show. In London, Hilda Leyel founded the highly successful chain of urban herbal emporiums, Culpeper's, developing along with her shop an archive of early modern and Old English herbal books and manuscripts that she made available to scholars and members. Through their engagement with urban culture and urban resources, women sought to assert their right to move from the country into the city and vice versa, increasing intimacy between both spaces and eroding the false binary that divided them.

Such urbanity, finally, has several important and disturbing impacts. Clearly, women at the top of hierarchies like the WAHI were often interested in prescribing change for rural places while assuming that their new methods were superior. Their burgeoning understandings of food justice clarify that urban ecologies are interwoven with rural ones as national ecologies are interwoven with world ecologies in an indivisible and sometimes disturbing mesh. At the same time, the often substantial privilege of women organizers and leaders in this movement, many of them influenced by eugenicist principles, certainly gave them a great deal of power in deciding what projects and people might be placed where, what crops should be grown and by whom, and what places and people mattered most. These top-down practices resemble, or perhaps helped to construct, the absence of local voices in large social movements such as environmentalism and conservation even now. As London acted as a gathering place from which experimental solutions and interventions emanated, it became a market and a forum for the exchange of ideas and material goods women brought into being on the land. This exchange was competitive, and the leaders were often ambitious, wealthy women, anxious to intervene in the Land Question and bring their new vision of a green world populated by New Women on "old acres" to life. They argued that "social movements spread from the top downwards" and assumed if "educated women" were leaders, "girls and women of the less educated classes" would follow suit.[93] Landswomen, herbalists, and market gardeners on the ground would have their own perspectives on their presumed superiority as early green leaders. They often simply ignored working-class women's knowledge or local resistance to agricultural work. In their inexperience they sometimes pulled up the wrong plants literally and figuratively, injuring both local people and a newly sacred entity landworker Rose Macaulay describes as "People's Food."[94] The meeting of the Land and Woman Questions then was of course marked by conflict, competition, and classism at every level. Their London meetings and publishers kept the voices of those at the top, like Lady Warwick, in print and in the news. Compelled by their own status, such women trusted in themselves as "introducers of order and method into realms where chaos ruled."[95]

Beyond Natural History

Finally, as elite women moved into agricultural and horticultural work, they used their privilege and credentials to shape conceptions of knowledge itself. Women like herbalist Maud Grieve or herbal historian Agnes Arber are among the first women to include their academic or professional memberships on the title pages of their books. Although they are primarily unsung, they are undoubtedly responsible for the dissemination and formation of the new science of ecology itself. Turning back to where this introduction began, with the example of Olive Cockerell's now very obscure work, it is possible to see how a paradigm shift from an epistemology such as natural history to ecology marks the published work of New Women.

Like many Victorians, Cockerell was an amateur naturalist and botanist from childhood. As an adolescent and art student, she developed a passion for fungi and collected it widely as a subject for her drawing and painting. However, by the early twentieth century, her status as a single middle-class woman of modest means meant that she left her identity as an amateur naturalist behind. She needed to professionalize. The census of 1901 records Cockerell as a "black and white artist" or illustrator, and even while providing childcare for her brother in distant New Mexico, Cockerell was sharpening her professional skills as a scientific illustrator.[96] One illustration that survives reveals that she pushed against not only the limits placed on women illustrators but the limits placed on an abstract Nature through Victorian understandings of natural history and botany. Even in scientific illustration, these clearly viewed the more-than-human world through a taxonomic lens. By contrast, the new science of ecology would examine dynamically "the relations between various organisms as well as the relations between organisms and their environments," focusing on what living things do among each other in addition to what they are.[97]

This new perspective is startlingly evident in Cockerell's extraordinary scientific illustration of the Southwestern American silk moth, *Oiketicus townsendi* (see fig. 2), now digitized for posterity in the collection of the international Biodiversity Heritage Library. A result of Cockerell's commission to illustrate a monograph on the silk moths of North America, it was published posthumously in 1913 in the journal the *Entomologist*, where it merits a full color plate as the first image of the desert moth and its life cycle.[98] Giving pride of place to the most striking stage of the moth's metamorphosis, the image features the moth's transformation into a spiny chrysalis that uncannily mimics the branches of the thorn tree upon which it pupates. A strange, armored body/object, the thorny chrysalis oozes acidic foam from its top, forming an escape hatch without which the emerging moth would be imprisoned in its own skin. Smaller representations of the moth in its other stages (larva, grub, caterpillar, and adult in flight) encircle this large cocoon, which is suspended from the tree branch, bisecting the page. Joined at the center of the page and at the intersection of ontological categories, the chrysalis and branch command attention. Their intimate entanglement invites the viewer to ask the central question of the then-new science of ecology—that is, where does the line between purportedly "inert" or nonagentic things like trees and living creatures like moths begin and end?[99]

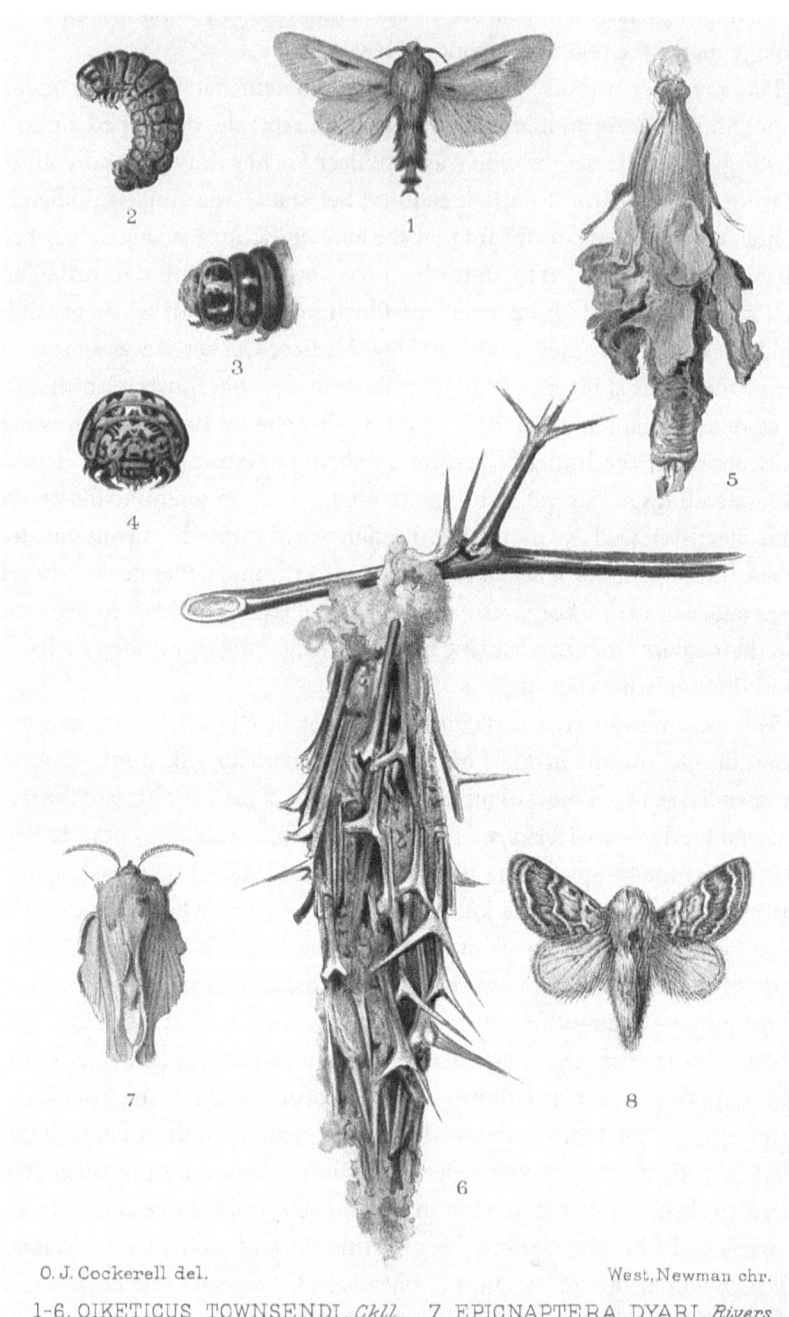

Figure 2. Oiketicus townsendi, Olive J. Cockerell, del., *The Entomologist,* March 1913, plate V. (Courtesy of the Ralph Brown Draughon Library, Auburn University)

Beneath this startling image, each stage of the moth's life cycle is dutifully labeled in its Latin name. That taxonomic gesture, however, seems almost inadequate in contrast to the illustration's central image of ontological anarchy, the material entanglement of what should be separately bound categories. The illustration prompts the viewer to consider each of the aspects of the new science of ecology: interdependence, the situated interactions of living things and abiotic conditions, the exuberance of life, its seemingly limitless diversity, the acute specificity of the material conditions of habitat, the phenomenon of cooperation among organisms, and niche construction. The inhospitable branch brings with it a glimpse into Darwin's entangled bank, a vision of a potentially hostile material world that shapes the background for the incessant "war of nature" and its laws.[100] However, the environmental story Cockerell tells in the image resists popular reductions of Darwin's theories that would make the moth's habitat a mere background to "the Struggle for Life" or the "law" of "Reproduction."[101] Instead, the image captures the moment of a creature becoming through its engagement with other life forms. As Cockerell's career evolved, so did her articulation of new central ecological questions raised by her depiction of the moth. These are visible in the visual and textual rhetoric of possibility that is clear in her published record of her own garden.

Her later work would, like her "*Oiketicus townsendi,*" stray far from familiar Victorian extinction narratives and explore the vibrant mesh between living things widely defined: people, the elements, and the plants, soil, insects, and animals in her raised-bed market garden. The illustrated garden text she would write with her partner Helen Nussey would include her own hands and image as proof of her capacity to transform, to become something or someone else in a new place, shifting from artist to gardener through her entanglement with the vibrant world around her. The very title of her book, A *French Garden in England,* engages the mobile spirit of New Woman ecologies. If the eminent Victorian George Eliot had famously used the regional dialect and voice of an elder to warn her heroine in *The Mill on the Floss* that "things out of nature niver thrive," Cockerell and her generation argue that there are as many Natures as there are definitions of what it means to thrive.[102]

Ironically, the creature Cockerell was commissioned to illustrate in New Mexico was, for the New Woman's antagonists such as the author Ouida, a "recurring metaphor" for the New Woman herself.[103] The moth dashing itself against a light or consuming fabric symbolizes the problem of the New Woman corroding the very fabric of life. Untrue to her own "essence" or biological determinism in her nonreproductivity, the mobile New Woman too was very likely to "ga[d] about," without constructive purpose.[104] In *Mobility and Modernity,* Wendy Parkins explains how this image of "a woman with freedom of movement signified a potentially unfettered female agency, which might pose a danger to the stability of the social and familial order."[105] Associating "processes of time and change with decay" in Ouida's novel *Moths* (1890), the hero complains that he "lives in 'a world of moths. Half the moths are burning themselves in feverish frailty, the other half are corroding and consuming all that they touch.'"[106] As Wendy Parkins argues, "By metaphor and metonym, then, women embody the excesses and moral decline of modern culture" through this image.[107] This purportedly destructive and self-destructive insect critiques the model of reproductivity and consumption upon which modern womanhood is constructed. In this discourse, the moth is always "less than" human, as the New Woman herself is less than a woman. The field of New Woman scholarship has gained much from reading such metaphors drawn from a hostile and essentializing Nature and the cultural work such living things—like flowers—do in characterizing women's mobility and modernity.[108] To be sure, the prevalence of such metaphors raises the stakes of women's environmental interventions. But the time is ripe for reading the New Woman's intersection with the materiality of such things themselves. In reading New Woman literature and culture, it's clear that narratives of the fin de siècle that place the figure at risk of decline and degeneration are countered by an often optimistic, sometimes troubled and troubling *aube-de-siècle* narrative interested in resilience, regeneration, and ecological futurity. Even during the darkest days of the Great War, Frances Warwick writes of "why [she] has so much hope for the future of the woman on the land."[109] Her voice displaces the narrative of conservation and decorative floral intervention associated with Victorians like Octavia Hill and enunciates an edible ecological futurity and a sense of the vibrancy of the material world that is based upon economic need and self-interest rather than beneficence.

To recover such New Woman ecologies, this book begins at the heart of the Arts and Crafts movement, with early green author Mary De Morgan and her exquisitely illustrated tales. Having become close to the inner circle of the Morris family through her skill in storytelling and embroidery, De Morgan expresses in narrative form what her close colleague May Morris and her sister-in-law, Evelyn Pickering De Morgan, expressed in embroidery and painting. As De Morgan critiques and displaces the ecological paradigm of the garden, her tales "The Rain Maiden" and "The Windfairies" (1900) develop an exhilarating alternative. They deploy the elements of wind, rain, and even earth as mobile, agentic forces that express, on the one hand, the vitality of the elements, and, on the other, the challenges of mobility and creative integrity represented by the image of the New Woman. Burning, raining, flooding, and blowing, these mobile elemental ecologies demand a disorienting acknowledgment of human interdependence, a recognition of elemental power, and the hope for strategic partnerships in making art for an ecological future.

Chapter 2 concerns De Morgan's illustrator, Olive Cockerell, who surprised her eminent godmother, Octavia Hill, by changing occupations entirely and becoming a market gardener. I trace how such women rejected monoculture to participate in the "third experience of alternative agriculture" in England in the early twentieth century.[110] In her garden, Cockerell and her partner Helen Nussey revived the traditional practice of "intensive culture," growing local food year round, in small spaces, using raised beds, in the traditional manner. This proves a counterpoint to the partners' use of mass-produced glass, soot fertilizer, white lead (which they used to mend broken bell glasses,) and train transportation (to exchange urban horse manure for their produce). I argue that, in their garden and their book, *A French Market Garden in England: A Record of a Year's Successes and Failures in Intensive Culture* (1909), the two apply the Arts and Crafts ideals of simplification and ecological interdependence to reinvent the material foundation of the rural household and shape a queer ecology based upon an ethic of partnership between professional women and the land.

Cockerell and Nussey clearly shape their garden as a kind of utopia, full of potential for imagining ecological futurity. Chapter 3 explores Utopian England more widely, examining the convergence of the image of the New Woman with the New Life, a transformative movement linked to the development of alternative communities such as Whiteway, in the Cotswolds. The attraction to these "little Utopias" is clear in the fiction and life writing of George Egerton, Edith Ellis (who wrote under her married name, Mrs. Havelock Ellis), Nellie Shaw, and Sarah Grand. They imagine cooperation, rather than competitive individualism, as the basis of evolutionary life. They engage the back-to-the-land ideology of Walt Whitman, William Morris, Peter Kropotkin, Thoreau, Tolstoy, Edward Carpenter, Havelock Ellis, and James Hinton. The future thought of these women often disrupts what José Muñoz terms "straight time," exploring a "queer relationality" that usefully intervenes in both individualist, masculinist environmentalisms and a heteronormative, ecocidal culture.[111] At the same time, the contribution of such women to often elitist movements such as conservation and preservation is an issue chapter 3 considers.

The Great War and the formation of the Women's Land Army disrupted some such rural utopias. Although historical propaganda and studies even now characterize women as "replacements" for male farm workers, in historical reality

Landswomen were offered the promise of permanent agricultural careers and a major role in the reinvigorating of English agriculture, which had languished for several generations. Chapter 4 explores how new recruits faced farmland that had gone fallow. In contrast to the heavily propagandized images and texts that placed the figure of the New Women in a pastoral England, land workers like militant suffragette Olive Hockin or recruits like poet Rose Macaulay found themselves spreading manure "like marmalade" in a desperate attempt to grow something in long uncultivated soil. Their task of transforming land and their own bodies, their new exposure to the elements, to labor, and to chemicals and machines that were to reshape them materially and conceptually, resulted in a push to view human and ecological well-being as one. Reading Macaulay's suite of "On the Land" poems as well as the memoirs, letters, propaganda, photographs, and ephemera published in the official journal of the Women's Land Army, the *Landswoman,* I explore ways in which recruits questioned their complicity with the war, encountered the same chemicals, fertilizers, and pesticides that were also used by and on soldiers abroad, and coped with the ironies of their work as growers who supported the war effort. This chapter examines the Landswomen's acute awareness of the ecological absence of boundaries on earth between matters and materialities themselves, as well as their resistance to the gendered and nationalized conceptual divides between them as the war progressed.

Survivors of the Great War, women herbalists are the subject of the book's fifth and final chapter. Well into the twentieth century, these women would create the new field of "modern" herbalism.[112] Led by herbal "luminaries" like Maud Grieve, entrepreneurs like Hilda Leyel, and scholars, writers, herbalists, and gardeners like Agnes Arber, Rosalind Northcote, and Eleanour Sinclair Rohde, these professionals formed a "circle of illustrious, knowledgeable women who were renowned and pioneering experts" in the field of herbalism in the 1920s and 1930s.[113] Like land workers, many of these women first professionalized during the Great War. From the beginning, however, herbalists were different. What was needed "to produce herbs successfully" was "a higher form of gardening."[114] During and after the war, Grieve and others led a new generation of women herbalists in this venture, first cultivating, harvesting, and processing potent medicinal herbs like belladonna, deadly nightshade, digitalis, and henbane, next seeking perfection in culinary and aromatic herbs like lavender. Herbalism then enabled the construction of a new area of expertise for women. Their labor, studies, and experiments led to their construction of a polymathic epistemology of herbs that included not just knowledge of plants but of early modern women's herbal histories in addition to, at times, an almost "euphoric" confidence in the role herbs might play in modern culture and the role women might play as experts of an "unquestionably" woman-dominated field.[115]

The sweetness of this coup d'état was intensified by the circle's ability to transform plants previously considered "inert" into vibrant participants in health, wellness, women's authority, and biodiversity.[116] However, just as women herbalists were to capitalize on their success, their work was curtailed and regulated by the passage of the 1933 Poisons Act and the 1941 Pharmacy and Medicines Act. These acts express the cultural anxiety around women's authority as caregivers and growers, an anxiety that is expressed in Agatha Christie's deployment of herbs as resilient—and potent—English plants in her detective fiction—for example, in "The Herb of Death" and *The Five Little Pigs*. While this final chapter of *New Woman Ecologies* ends with a discussion of the Pharmacy and Medicines Act of 1941, which strove to drive women herbalists out of business, the epilogue follows the work of historical New Women in preserving the public memory of their ecologies for posterity. The epilogue begins with a last act, the donation of a gendered herbal object, an early nineteenth-century herbstrewer's dress, to the Brighton Pavilion Textile Museum in 1961. Such memory keeping preserves a dual record of resilience, recording women's resistance to a modernity fatal to ecology and their resistance to the entangled projects of monoculture and women's domestication during and after the Second World War. Their resistance to the latter is timelier than ever.

Indeed, in their recent study *Biodiversity, Ecosystem Functioning, and Human Wellbeing*, Shahid Naeem and Daniel Bunker critique "conventional approaches to ecology," as these "often lack the necessary integration to make a compelling case for the critical importance of biodiversity and human wellbeing."[117] The science and movement they term "pure ecology" often focuses on "pollution, the ozone hole, climate change, collapsing fisheries, disappearing forests, the adverse consequences of unbounded human population growth, emerging diseases and conservation biology; this last topic being where the value of biodiversity dominates."[118] "This approach," they argue, "obscures the inextricable links between biodiversity, ecosystems, and human wellbeing."[119] The exclusive concerns of "pure ecology" do not "prepare one for understanding and applying ecology in the context of the modern world," where we must ask "the question that is rarely asked by ecology texts—*What is the significance of biodiversity to human wellbeing?*"[120] Nature writing, traditionally defined in English literature, has often mirrored the position of pure ecology in science studies, seeking to represent remote areas, mountaintops, and ruins that stress the absence of most humans other than the usually male writer. The reclamation of situated, historical, gendered ecologies such as New Woman ecologies contests that model through its close adherence to this new ecological imperative. Alternatively, if the New Woman's ecological story fails to be understood as such, we lose an extant example of how women's situated knowledge matters to ecology, a field that can no longer sustain the idea of an abstract Nature dwelling apart from any particular human culture. Inspired by Donna Haraway's feminist claim that "the only way to find a larger vision is to be somewhere in particular," this project follows a specific community of women in their historical moment, identifies patterns in their experiences and representations, and notes that their alignment with modernity is generative, resulting in new partnerships between the material, elemental world and people.[121]

Indeed, if nature writing once explored the great luxury of edging people out of the picture, we now face a global ecology in which an agentic earth and its elements do that for us, placing the most vulnerable human populations in danger. These elemental conditions, moreover, may be understood as acting in response to our food ways, to the damage done by outsourcing and transporting food, to the deforestation, drought, heat, flooding, and toxic runoff that is the result of monoculture. A world without people, once a utopian fantasy, has now become a distinct dystopian possibility, as we lose ground literally to rising waters, drought, and severe weather events that may make places like the Lake District, as Sarah Hall darkly imagines in her *Daughters of the North* (2008), eventually uninhabitable. In returning people to ecology now, it is instructive to revisit historical precedents. Barbara Gates paved the way for this work with her broad and deep study of the altruism of Victorian and Edwardian women who, on the basis of an ecofeminist "creed of kinship," sought to "nurture nature," expand botanical knowledge, and defend birds, urban land, and animals from the depradations of modernity.[122] Under particular circumstances, however, women of the same time period turned to modernity itself to answer both the Woman and the Land Question affirmatively. Studying the ecologies of the New Woman results in witnessing both historical mistakes and innovative problem-solving. It means revisiting the imagination of an ecological modernity and the practices designed to create it which seem, in the twenty-first century, suddenly and ironically, fresh.

1

ELEMENTAL ECOLOGIES

Arts and Crafts Women and Early Green Thought

The tales of Mary De Morgan are a prime example of the intersection between New Woman literature and the transformative early green culture of the Arts and Crafts movement. Easily fitting the image of the New Woman herself, De Morgan was a member of the Women's Suffrage League, an author, a journalist, an embroiderer, a typist, a secretary, an East End volunteer, and a prison reformer.[1] Through her skill at storytelling and embroidery, she lived and worked at the heart of the Arts and Crafts movement, becoming close to the Morris and Burne-Jones families. One of her four collections of fairy tales is dedicated to the grandchildren of the latter, and each collection is lavishly illustrated by central figures within the movement: her brother, William De Morgan, Walter Crane, and Olive Cockerell. Like the fairy-tale collections of Charles Dickens and Oscar Wilde, the four collections were published as Christmas gift books and were once considered by reviewers to be among the most "original" and "most elegant giftbooks" they had ever seen.[2] Welcomed as "real treasure[s]" or "gem[s]" in journals like *Punch*, De Morgan's tales are now beginning to be recognized as innovative narratives that utilize the strategies of the New Woman writer to critique both the subjection of women under Victorian marriage laws and industrialized capitalism.[3] Hybrid objects of both New Woman and Arts and Crafts culture, De Morgan's books combine the former's performative narratives and visual strategies, her avowed feminist politics, and the deliberate archaisms and early green sentiments of Arts and Crafts. The mixture brings a number of vital bodies into view in unexpected ways, resulting in "beautiful and wicked princess[es]," terrifying enchanted rosebushes, and elemental dancers constructed of wind and rain.[4] Such beings strike a keynote for New Woman ecologies. They disrupt the Victorian ecological paradigm of the garden queen in the flower garden and create a strategic alliance between the mobile image of the New Woman and a vibrant elemental ecology.

De Morgan's affinity for the elements greens the anarchic mobility of the New Woman and genders the early green ideology of Arts and Crafts. In tales such as "The Windfairies," "The Toy Princess," and "The Rain Maiden," De Morgan naturalizes women's bodily autonomy and artistic integrity while she critiques the joint commodification of women on the marriage market and industrialist-capitalist projects such as mass production and mechanism. Alternatively, in tales like "Handa and Siegfried" she models cooperative labor and artisanship in the tradition of the early greens, such as her close associate William Morris and his colleague the anarcho-communist alternative agriculturalist Peter Kropotkin, about whom she wrote in the articles "Co-operation in England" (1890) and "The New Trades-Unionism and Socialism in England" (1891). There she considers anarchist communes as alternatives to commodity culture and lauds the ideal of people "shar[ing] the result of their labor."[5] The latter ideal, she writes, although difficult to achieve in practice, "gains dignity by having for its leader Prince Peter Kropotkin."[6] He "must in every way command respect . . . in his private and public character, as a scientific man."[7] De Morgan's praise references Kropotkin's reenvisioning of Darwin's political and ecological legacy. While many had naturalized capitalism via Darwin's emphasis on competition and struggle in *The Origin of the Species,* Kropotkin would ultimately naturalize communism through Darwin's equally compelling vision of mutualism. The latter became an important component of late Victorian and early twentieth-century thought, evident in New Life utopian experiments in communal living.[8] De Morgan too seems intrigued by the idea of mutualism, cooperation across and between ontological categories. But her interest in the transformative culture of the early greens is made her own in her imaginative tales. In these fictions, her critique of competition and industrialization is matched by the power of more-than-human things, especially as they aid creativity in the tradition of Arts and Crafts.

As the classical and early modern elements of earth, wind, fire, and water are the "animated materialities with and through which life thrives," they became the very stuff of that movement's deliberately archaic techniques.[9] Pots and tiles were "fired," paints were composed of precious metals from the earth, air was crucial to a just working environment, and artisans hammered iron into useful and beautiful things. De Morgan the writer works in that tradition too, exploring the rich materiality of the elements. Animating the air through the characters of "The Windfairies," or water through the figure of "The Rain Maiden," De Morgan establishes an intimacy and interdependence between human and more-than-human things that fosters mutual respect rather than an anthropocentric ethic of stewardship. Her early critique of plant collection,

"The Pool and the Tree," for example, creates water as a willful, desirous, and mobile thing, while the human plant collectors and gardeners in the tale are characterized as thieves and fools. De Morgan then constructs horizontal relationships in which humans, rain, wind, and clouds come into being through each other. Interdependence and cooperation, or what Peter Kropotkin described as "mutual aid," among living things rather than competition is the theme of these tales.[10] Dancers in "The Windfairies" and "The Rain Maiden" form strategic alliances with the elements in De Morgan's ecology, naturalizing the New Woman themes of mobility, sexual and artistic integrity, and vocation itself. They come to recognize, in the process, the sometimes volatile generative powers and exuberance of the elements in narratives that stand quite apart from the more familiar Victorian fairy-tale fare. Ultimately De Morgan explores the intrinsic value of living things—women, plants, rain, clouds, and wind—resisting their commodification on the market or their instrumentalization in industry. This insight follows her analysis of a less appealing feature of Arts and Crafts culture: its near-obsession with the gendered ecological paradigm of the garden.[11] De Morgan begins a powerful critique of this paradigm early on with the dark tale "The Seeds of Love," a thorough and damning critique of what Amy King terms the "botanical vernacular" in which "blooming and marriageability coalesce."[12] This narrative is endemic to the realist novel, the visual culture of Arts and Crafts, and mainstream Victorian culture itself. The fairy-tale form, then, allows a less "decorous" treatment of that discourse, exposing its hidden violence and objectification of women.[13]

Damned to Garden: "The Seeds of Love"

In "The Seeds of Love," from her second collection *On a Pincushion and Other Tales*, De Morgan critiques the appeal of the garden and the familiar, domesticating discourse of the garden queen.[14] She knew these discourses in her very hands as an embroiderer. In that craft as well as in the artwork of Arts and Crafts, the garden epitomized the movement's concept of a green utopia often problematically inhabited by an objectified woman. The latter intersection is clear in the consistent objectification of De Morgan's friend Jane Morris in floral paintings like *The Day Dream* (1880). As Jan Marsh aptly notes, this painting is not a portrait; rather, Morris's very identity is subsumed in the work "into the spirit of Nature."[15] Through his considerable skill with composition, brush strokes, and color, Dante Gabriel Rossetti successfully blends the human woman with the goddess Flora and conflates both with the tree, the flower she holds, and the season. The ephemeral quality of each is accented by the glow of the red-tongued honeysuckle bloom at the center of the painting that seems to express Morris's "bloom" or her perceived sexual availability itself.[16]

Arts and Crafts women artists themselves were constantly subjected to such visual iconography, which mirrored the "rapprochement between women and flowers" that was rampant in Victorian mainstream culture as well as in bohemian schools.[17] In the static portrayals of women in pre-Raphaelite paintings, their mobility is sharply curtailed, reduced to a pose. This is clear in the work of Dante Gabriel Rossetti, William Morris, Frederic Leighton, Edward Burne-Jones, and others. Certainly, this conflation of women and plants or flowers is visible in the embroideries to which De Morgan contributed her labor, executing designs by both Morris and Burne-Jones. In such textiles, images of women harvesting mirror their own fecundity rather than explore their productivity. A famous tapestry of Pomona, for example, shows her holding apples in her lap rather than picking them. This visual culture suggests the extent to which the school of Arts and Crafts accommodated quite standard Victorian gender norms, such as the conventional idea that "the ideal woman is one who is close to nature, practicing her role in life by working with flowers" or fruit.[18]

Edenic scenes featuring women at this kind of work signal the triumph of the Morrissian postrevolutionary utopian society in which work becomes a pleasure. But this achievement is often signified in visual or literary terms that make women's reproductive role a given. This has the effect Patricia Murphy identifies in the Victorian period as "trapping the female in time," always slightly behind the evolution of men.[19] In texts and tapestries, women's time is also undoubtedly heteronormative or straight time; women are represented awaiting what Morris imagines as the ecstatic arrival of "real marriage," in which women are finally "free" to select a heteronormative partner regardless of financial status and without the tyranny of the period's marriage laws which deprive women of agency.[20] In this new utopia, figures of Pomona or Flora, the goddesses of harvest and spring, offer themselves to the viewer, seeming to speak for all women: "I am the ancient apple queen, as once I was so am I now"; "I am the handmaid of the earth."[21] Such representations of women, as in *Tess of the d'Urbervilles*, "perpetuate the notion of an unchanging female essence across the ages and negate the possibility of substantive improvement in women's status."[22] Martin Delvaux notes that Morris "developed Marx's anti-capitalist critique, linking it to questions of gender and transferring it to an environmental level."[23] However, Jan Marsh notes that "heterosexuality rules"[24] in Morris's utopian garden states and that his stipulation that an authentic marriage is a matter of simple "inclination" between men and women reveals that he can imagine no other alternative sexualities or life partnerships.[25] The very basis of Morris's early green vision of the future as expressed in "The Dawn of a New Epoch" (1886) is of the nation itself as a gendered garden. In *News from Nowhere,* a transformed London is "running over with flowers" and populated by "young girls" offering Roger and Richard Hammond fruit and blossoms.[26] The women who worked for Morris and Co. would have met this reproductive and heteronormative floral discourse as it manifested itself in the Morris house designs that they executed.

Indeed, as embroiderers, De Morgan and May and Jane Morris, were, in a sense, expert gardeners.[27] Skilled in the floral vocabulary of Morris's designs, the three women worked together on the intricate hangings for Morris's bed at Kelmscott Manor, interweaving native English meadow flowers on the bed's coverlet and a trellis of roses and birds on the curtains, with the text of Morris's poetry crowning the top as a border and a ribbon of blue, representing the Thames River, coursing around the bottom of the coverlet.[28] The skill of their embroidery is unquestionable; however, it was unsigned and often, although not always, executed men's designs. As Anthea Callen has written, a "lack of any unselfconscious integration of men and women" emanated from the heart of the Arts and Crafts movement outward to all its levels, compromising its radical aims and becoming "reactionary in its reinforcement of the traditional patriarchal structure which dominated contemporary society."[29] Although the movement believed that self-expression through the making of art and craft was a basic human right, women's positions in the movement were often subordinate to those of men. The progressive ideology of even Morris's futurist vision of the nation as a garden state cannot overcome the gendered mythologies of that space. De Morgan was well known to have argued about socialism with William Morris. Although details of these debates are elusive,[30] in her published writing, she clearly critiques the sexism of "English socialists," pointing out that "many of their leaders assum[e] women's place in creation must be secondary, and that their real emancipation would be injurious to humanity."[31] In her literary fairy tales, when De Morgan imagines the injurious quality of sexism and the exhilarating experience of "real emancipation," she starts by questioning the garden and its queens and ends by indicting both as unsustainable options. In their place she creates original alternative myths that mobilize rather than sequester women.

De Morgan's interrogation of the garden as a gendered ecological paradigm is clear in tales like "The Seeds of Love." Enclosed within a narrative frame that gives voice to a woman's most intimate objects—a jet shawl-pin, a "cut and polished" pebble brooch, and a common pin—"The Seeds of Love," which is narrated by the black jet shawl-pin, is a sad tale of a white queen, Blanchelys, damned to garden. The jet pin, ominously associated with mourning jewelry, begins "The Seeds of Love" by depicting two young girls, cousins, who have the dull task of mechanically opening and shutting a bridge gate. As the river rushes below and traffic flows above, the girls remain in one place. Their grandmother does nothing but sit by the fire and knit as the girls work. Blonde Blanchelys, the virtuous cousin, tolerates the monotonous work well, while dark Zaire grumbles. Opportunity strikes, however, when, on the eve of her death, their grandmother gives each girl a magic candle that, when lit in the dead of night, will produce a fairy "who will grant you the wish of your heart. If it is a good wish it will be a good fairy that appears, but if it is a wicked wish it will be a wicked fairy that comes."[32] Blanchelys lights her candle and conjures a boy fairy, personifying Love. She asks him to "give me the love of the King's son," and he complies, shaking several seeds from the branch of roses he carries into the young woman's hands. These seem potent entities from the start: "More like jewels than seeds . . . bright clear red, like rubies, and each one was in the form of a heart." "Take them . . . and plant them in the earth," the boy says. "From them will spring a rose-tree, and as it grows so his love for you will grow. While that tree lives he will love you more than all the world, but should it pine and die his love for you would wane and die also."[33] Blanchelys kisses them and places them in a flower pot in her room: "'Now I can watch it both night and day,' she said, 'and see that no harm comes to it.'"[34] The first time Blanchelys speaks to the King's son, the plant puts forth "a tiny shoot." As the lovers converse, "the plant in the pot grew larger and larger, till at last . . . it was a rose tree . . . covered with tiny buds."[35]

By the time the prince proposes, the "buds had burst into a splendid white rose, which scented the whole room." After she is married, Blanchely's plant grows "into a big tree" and she plants it in the palace garden. Her "first care in the morning was to examine her rose-tree, whilst she brushed from it all insects, and cut off the dead leaves." But as her cousin gardens, Zaire's jealousy grows. She thinks of "nothing but how she could kill the rose tree. . . . First she pulled off its leaves, and cut its branches, but fresh leaves grew. . . . Then she took a sharp knife, and pierced it through the trunk, and peeled off the bark, so that it bled . . . but the tree grew and grew."[36] Soon Zaire too lights her candle, and a fairy appears, who is Envy personified. She gives Zaire a glittering green snake and tells her to dig to the roots of the tree and place it there. In an elegant collusion of text and image, De Morgan's text wraps around a serpentine illustration by her brother William (see fig. 3); both underscore Zaire's exquisitely wicked reversal of Ruskin's garden queen, who epitomizes

Next day when Queen Blanchelys came to look at her tree she found it drooping, so she called to the gardeners to give it water, but not all the water in the world could refresh it, and each day it drooped more and more, and the flowers began to die and fall away. Poor Queen Blanchelys watched it with tears in her eyes. She sent for gardeners from far and near, but they could do nothing for it, and the Queen was sick at heart, and grew pale and thin, for she knew that her husband was beginning to love her less and less.

Figure 3. "The Seeds of Love," from Mary De Morgan's *On a Pincushion,* 1877. (Courtesy of the De Morgan Foundation)

domestic expertise as well as "philanthropy."[37] In the drawing, her body itself is a serpentine form that begins with the tree's exposed roots, on which she stands with her pursed lips, which point upwards, ready to kiss the snake for good luck. Several sharp points, including her toe and her garden spade, underscore the pointed criminality of her digging. The snake, shockingly coiled around one hand, and the shovel in the other mark Zaire as a woman who takes the culture of bloom into her own hands.

At this point, the women's placement in a discursive, raced plot is solidified. The garden becomes a map of the world that ironically restricts rather than expands their possibilities, categorizing them as either mistress or slave. Naming Zaire after a rebellious colony in the Congo, De Morgan constructs Blanchelys's virtue through the abjection of a racial other. Zaire is "bitterly jealous" of her sister, "though Blanchelys was kind to her and gave her beautiful things, and took her to live with her at the palace."[38] In this economy, Zaire is allotted only one desire: to envy what her cousin has. Her digging and destruction of the rose at the roots reverses postcolonial ecologies, bringing ecological damage and destruction home. At the same time, Blanchelys's gardening and Zaire's attack on the rosebush show their intimate connection to the limited options of each of their garden plots. Their opposition is grown from the same discursive soil.

As an invitation to horticultural disaster, the high-maintenance white rose tree represents the unsustainability of the white garden queen. It is clear from the beginning that the interlocking of the rose and the marriage is a risky joint venture; the suspense of the story lies in awaiting the rose's inevitable death at the hands of a much more ambitious and arguably more interesting, less subservient woman character, whose sheer, shocking rebelliousness, wickedness, criminality and physical autonomy makes up one of the great pleasures of the tale. If representations of gardening are powerfully linked to the formation of identities in Victorian culture, through the gendering and naturalizing of everyday practices such as cultivating, collecting, planting, and digging, then it becomes clear that in many ways Zaire has been given the more satisfying job, the "heavy" work of digging to the roots of the plant to kill it.[39] Ironically, in defiance of the gendered tradition of Victorian garden writing, Blanchelys's light gardening—the constant pruning, watering, and tying up of the rose— evokes great anxiety and tension as the reader awaits the inevitable failure of her efforts.

De Morgan's garden tale critiques the mystification of marriage as natural or pleasurable and characterizes it as constant work on the part of the wife— just so that she might stay in place. As Zaire digs, she strikes at the root of the problem of bloom: its short life. As the snake constricts the rose's roots, the image suggests the narrowing of the good sister's options and plots. Carrying the little green snake in her bodice, raising it up to kiss it with her competent arms and hands, and planting it at the roots of the tree, Zaire chooses another path. She is at once shockingly criminal and, in the tradition of linking the alterity of the ambitious, glamorous New Woman to criminality in general, she is delightfully transgressive.[40] William and Mary De Morgan's joint representation of Zaire's destruction of the rose in text and image maximizes the future shock of that image, not least by featuring her good-luck kiss to the snake. Clearly, this remyths Eve's shame in the garden. It also captures what Elizabeth Miller describes as a tradition in the representation of the glamorous New Woman as criminal, the "*pleasure* of such transgression[s]."[41] As contemporary reviews of De Morgan's work suggest, readers delighted in these wicked and beautiful characters. The aptly named Blanchelys, and her plot of ephemeral bloom, martyrdom, and self-sacrifice, pales in comparison to the picture of Zaire.

When the rose tree dies—and with it the king's love—Zaire takes her cousin's place. Queen Blanchelys goes to look for Love, only to find him at a funeral, where a woman mourns the abusive husband who had "made her work for him day and night and never gave her a kind word."[42] When she asks Love to tell her how to win the King's love back, she learns that her past gardening is just a precursor to the greater sacrifice to come. Love instructs her that the only way to recoup her love is to pierce herself with the thorns of the dead rose tree until she dies. As she does so, the tree appears "covered with green leaves and rosebuds." When "the rose-tree burst into bloom," however, the white roses become "red as the blood which sprang from the Queen's heart."[43] The king regains his love for Blanchelys and exiles Zaire, but, of course, nothing can bring the gardening queen back in human form again. Her transformation is complete.

The jet shawl-pin who narrates "The Seeds of Love" is proud of the aesthetic effect of this ending, the transformation of the garden queen into the rose, the queen of flowers, but the others on the pincushion weep. The reader too is left wondering if the cost of maintaining—and becoming—the rose tree is not too much and if, in fact, a system of social ascension through marriage makes women, like the rose tree, vulnerable to a host of problems.[44] While gardening is often represented as a form of women's creativity in the Victorian period, Blanchelys's rose tree is merely a type of domestic maintenance. A questionable gift, it soon becomes her mistress and she its minion. The tale's dramatic palette of black, white, red, and green—clear in the women's names, the rose in its two incarnations, the plant's vivid red heart-shaped seeds, and in the green snake—construct a dangerous slippage between images of a benign Nature and ecogothic horror. Blanchelys's very name becomes her fate by the end of the tale as her blood flows into the plant, reviving it and leaving her in death a white lady indeed.

The brooches, pins, and other women's objects who listen to this garden story while literally stuck in a lady's pincushion underscore its melancholy resonance as a tale of woman's enclosure. "'For my part, I like stories that end up all right,' remarks the Pebble Brooch." The common pin on the other hand "could not speak, for it was crying quietly, and was dreadfully ashamed of its tears. . . . Even the Bracelets had stopped chattering, and come down to listen. . . . The Shawl-pin, however, smiled. He felt his story had been a success, so he did not mind what the Brooch said."[45] The narrator's smile and his self-satisfaction resonate. Is his pride in his work cruel? At what cost is his aesthetic "success" achieved? The reactions of the audience on the dresser invite the reader to question the trajectory of such an aesthetic triumph, the achievement of the ultimate rapprochement between tenor and vehicle, the exchange of a woman for a flower.

De Morgan's tale speaks back to the representational currency of a gendered culture of bloom. While she engages the early green utopia of the garden and gardening, she also undercuts the power of the garden queens and cultivating women who populate the Arts and Crafts movement and the mainstream Victorian imaginary alike. In contrast, the spectrum of gently unconventional to sometimes terrifying women characters De Morgan creates from Princess Ursula to Zaire to Princess Fiorimonde celebrate, like the New Woman criminal or aesthete more generally, a fresh pleasure in alternative plots and characters.[46] When De Morgan writes *The Windfairies* in 1900, she becomes even more critical of garden enclosure and more supportive of rebellious women, ultimately abandoning the garden to create elemental ecologies that explore the pains and pleasures of the transgression of the gendered limits of such spaces. In this alternative ecology, De Morgan shifts the definition of Nature from the garden as a retreat or a place apart to a mobile thing at once within and without the limits of the human.

Elemental Ecologies

Scudding, blowing, raining, and flooding, Mary De Morgan's elemental ecologies in her later collection *The Windfairies* pick up where her imaginative critique of the garden ends. Tales like "The Rain Maiden," "The Windfairies," and "The Pool and the Tree" enliven things usually considered inert or mere resources in modernity. In a reversal of the latter, wind resists containment and raindrops knock insistently "to gain admittance" to domestic spaces in De

Morgan's tales.[47] Agentic figures, the elements save a "maiden" from marriage or gift her with dance. Insisting upon ontological intimacies, they are cut loose from reproductive models and resist all types of commodification. As humans engage the elements in De Morgan's tales, they learn that the elements are generative, becoming themselves through their encounter with other things. As water transforms into cloud, bodies seem mutable and changeable, full of possibility, and unpredictable, like the weather. Humans are less autonomous from, more entangled with, and dependent upon such powerful actants, which normally kindle cozily on the hearth or pour docilely from the tap. The tales "The Pool and the Tree," "The Rainmaiden," and "The Windfairies" develop both the relational and intrinsic value of elemental things, anticipating the first premise of conservation that is often clear in Arts and Crafts ideals.[48] At the same time, *The Windfairies*, like "The Seeds of Love," links the power of the elements to the anarchic qualities of the New Woman figure.

Arts and Crafts elemental literature and material culture celebrates the classical elements of earth, wind, fire, and water as an alternative ecological and aesthetic paradigm, foundational to life and art. This nineteenth-century school of thought anticipates the ecological benefits of thinking elementally that have recently appeared in collections such as *Elemental Ecocriticism*, edited by early modern scholars Jeffrey Jerome Cohen and Lowell Duckert. That work argues that the medieval and classical cosmology of the elements models a welcome paradigm that goes beyond green. "Smaller than Nature," the elements are the "perceivable foundations of which worlds are composed, the animated materialities with and through which life thrives."[49] Because we are composed of water, breathe the air, and eat from the earth, the elements usefully erase the slash in nature/culture. "They are the outside that is already within, the very stuff of cosmos, home, body, and story."[50] Thinking elementally is useful to us—and, clearly, to Victorians like William Morris or Mary De Morgan—because in a time of darkening or dangerously sunny skies and rising waters, the elements remind us that they remain agentic, "inherently creative," as well as destructive, "motile . . . never inert."[51] The elements' response to human industry was clearly evident to Ruskin in his great lecture "Storm Clouds of the Nineteenth Century," as Jesse Oak Taylor notes.[52] Thinking elementally, Ruskin and William Morris could not help but express concern about the air materially transforming into a thick thing of our human "manufacture."[53]

The elements also demand respect in the school of Arts and Crafts because they are the craftsman's willful partner. William Morris writes in "The Art of the People" that Britain had forgotten the fields smiling, "the exultation of the fire," and the "countless laughter of the sea."[54] This elemental sovereignty, however, could be regained, touched through materials, through craftsmanship, and even through work habits. Hence, fire powered William De Morgan's kiln and open windows and fresh air supported the work in May Morris's embroidery studios. In the work of Mary De Morgan's sister-in-law, painter Evelyn Pickering De Morgan, both *Storm Spirits* (1900) and *Daughters of the Air* (1905) represent the transformative potential of women's relationality and autonomy through motile air. Sir Frederic Leighton's "Flaming June" or May Morris's airy designs like "Honeysuckle" celebrate ontological intimacies between the elements and human or plant life. Likewise, the element of air is valorized in Ruskin's tale "The King of the Golden River" (1843), as the trumpet-nosed personification of the Southwest Wind, Esquire, takes revenge for the mining of Treasure Valley. The image and sound of a wind instrument resonating through free air is a call to the ultimate transformative movement, revolution, in the emblematic title of the socialist journal *Clarion*.

As the elemental stuff of early green art and craft, wind and rain provide Mary De Morgan with an important alternative to the heavily gendered Arts and Crafts image of the garden. As Cohen and Duckert argue, the elements embrace difference, the entanglement of all life on earth, and the power to destroy and "regenerate," unhinged from heteronormative reproduction.[55] The elements were in fact a daily reality in the household Mary De Morgan once shared with her brother, potter William De Morgan, whose kiln literally blew the roof off of their house as he labored to reclaim the lost art of lusterware.[56] Mary De Morgan's friend and associate May Morris has discussed how her visits to the De Morgan household were spent "wandering through the house and garden eager over the latest" such "experiment" in the kiln. Unlike harnessing water, wind, coal, or fire to make identical "art manufactures," William De Morgan represented the mutability of human and elemental agency in each volatile glaze he mixed.[57] Each pot bears signs of a haptic entanglement with the elements in motion at a particular moment. May Morris writes that each firing was different, "a pot that had roused no special expectation came out a triumph of shining color amongst the ruin of a whole firing; there were 'spoilt' pieces that one could not help loving for some special quality in them—in short a whole chapter of the story which, passing under the eyes of those familiar with the building up of a craft, was alive with incidents hailed and followed with keenest interest."[58] The explosion of the kiln, the "ruin of a firing," as well as the "triumph of a color" that emerges from the fire may each provoke the power of the elements and the experience, in success and failure, of encountering matter May Morris describes as "alive." The vitality Morris communicates in this memory expresses her community's fascination with an elemental creative process that could rival and exceed the machine work of the industrialized world. This was the case in the triumphant end result of William De Morgan's labors, the "Moonlight Suite," in which earthen vessels and pots are fired with cobalt blue and luminous metals, depicting an ocean of waves powering ships in full sail.

Elemental intimacy, then, was one way in the school of Arts and Crafts to get "back to nature," a process facilitated by "breaking down the barriers between human beings, the sexes, and humans and animals."[59] In engaging the elements as well, the artist realized that "his first efforts at civilization came from Mother Earth, whose son he believed himself to be and his ashes or his bones returned to Earth enshrined in the fictile vases he created from their common clay . . . begun in the simplest fashion, fashioned by the simplest means, created from the commonest materials."[60] This elemental language, charting the cycle of human life through its intimacy with clay, aptly reveals the intimacy between nature and culture in the Arts and Crafts creative process. Such closeness modeled a utopian possibility for a partnership between the things of the earth and human artistic endeavor. The latter would bring out the best in both, in contrast to industrialization, which was dystopic, bringing out the worst in the same intersection. By 1889, Edward Carpenter, a student of William Morris, would write that "as man's power over materials increases he creates for himself a sphere and an environment of his own, in some sense apart and different from the great elemental world of the winds and the waves. . . . He creates what we call the artificial life . . . and, shutting himself up . . . shuts Nature out."[61] Alternatively, art should "stand in close relation to the earth and the sky and . . . elemental things, admitting no gulf between themselves and them, but only perfecting their expressiveness and beauty." This would be "true Art."[62] In practice, craftspeople established a commonality with and interdependence upon elemental things. However, in this description of the integrity of the creative process, the craftsman is decidedly male and the material itself is decidedly female, "Mother Earth."

How could women engage this elemental aesthetic without risking their own objectification? The answer would seem to be clear in a number of works by Arts and Crafts women. May Morris's innovative designs for Morris and Co., for example, are widely credited for creating an "airy" aesthetic, shifting the dark tonal palette of the firm's house style by replacing her father's dark backgrounds with negative space. In a design like "Honeysuckle," the effect clearly makes air visible.[63] But Morris's "airy" designs are more than simply a new aesthetic. They follow logically out of her political beliefs and work practices, intertwining with her discussions of crafts people's dependence on clean air to do their best work. Often photographed herself at work before an open window, she encouraged her students to do the same, not just for light, but for air, to oppose the stultifying practices of women's work in sweatshops. "The girls who sit in a stuffy workroom" to embroider by hand, she argued, "would be little or no worse off than if they worked a sewing machine all day."[64] For May Morris, access to fresh air was crucial to the process of making art with integrity, and it was crucial that art with integrity represented fresh air. The value of the latter might be represented even in the urban home, where the negative space of designs like her wallpaper "Honeysuckle" represent clean air, countering the smog of London, creating a utopian space already washed clean, reflecting the revolution to come. Likewise, May Morris's pillow embroidery of a pomegranate tree, floating upon a deep blue fabric background, suggests her fascination with the interdependence of plants and air. Her "Honeysuckle" wallpaper and other airy designs like it contrast enough with the reality of London skies during May Morris's life to suggest that they envision a postrevolutionary sky, a utopian ecology.

De Morgan's sister-in-law Evelyn Pickering De Morgan, like May Morris, also anticipates a coming upheaval and an equitable future in her later works, like *The Storm Spirits* and *Daughters of the Air*. These are deeply engaged in constructing alternative elemental allegories, representing the elements of air and water as sober, dark, and powerful feminist female figures. Such representations are a welcome counter to women's immersion in the language of flowers, challenging the exclusive masculinism of early green culture and the masculinist "brotherhood" of Arts and Crafts with an elemental sisterhood. Pickering De Morgan makes use of the elements to represent, again, bodily autonomy and mobility in her "sisterhood of women" that "portray[s] energy and action rather than constraint or passivity."[65] As Elise Lawton Smith argues, the elements appealed to Pickering De Morgan because, unlike land which can be "tilled, planted, domesticated, the forces of water and air have typically been seen as free from such control."[66] The airy effect of Morris's "Honeysuckle" is likewise achieved through its free suspension of the plants in air. In both cases the elements "serve as an effective allegory of release, both from the fetters of human embodiment and from imprisonment within socially defined roles."[67] Smith points out that both *The Storm Spirits* and *Daughters of the Air* thwart "the standard outcome of a postulated female/nature bond which has been to see the natural as incapable of constructive action and therefore dangerous to civilized stability."[68] De Morgan's elemental figures do not, like the works of the Pre-Raphaelite brotherhood, seek to seduce men but rather to move women toward a similarly potent sisterhood.[69]

The feminist allegorical representation of the elements in Pickering De Morgan's work, Elise Smith notes, clearly diverges from mainstream masculine representations of women and the elements that are often excuses to paint women as prone, nude, or in diaphanous drapery. To oppose this sexual objectification of women, Pickering De Morgan's stately figures like "Flood" in *The Storm Spirits* are cloaked and solemn. Flood steadily pours water from an urn into the ocean while "Cloud floats by at the top of the painting in the mysterious darkness of intricately folded drapery and feathered wing."[70] "Lightning" is draped in bright red and "poised for flight, her right hand already sparking with lightning bolts."[71] The "strength of mind and muscle required of such powerful spirits" creates a very different message, Smith argues, from pre-Raphaelite paintings of mermaids and watery femme fatales or Arts and Crafts Floras and Pomonas. Instead Pickering De Morgan's paintings represent "water in its strongest, freest form;" their "sisterhood suggests communal strength."[72] If New Woman art and literature deployed the idea of hybridity to inhabit "the spaces between old, fixed categories," Pickering De Morgan fuses women's bodies with archaic elemental forms to allegorize their autonomy.[73]

Mary De Morgan also explores such hybrid elemental identities in her own elemental representations. However, rather than deploying the elements as allegorical figures, she establishes the alterity and mobility of the elements and the female artist side by side. Through speaking trees and bodies of water, dancers, fairies, and elemental hybrids, the community De Morgan assembles may be anthropomorphic in the tradition of the literary fairy tale, and this tendency may depreciate her ecological work. However, at the same time, "a bit of anthropomorphism" between bodies may usefully erode the distance between culture and nature, drawing people and all sorts of vibrant bodies conceptually closer together as linked beings, interdependent and entangled.[74] Finally, whereas Pickering De Morgan's draped elemental women are often quite stately and dignified, clearly refuting the sensuality of earlier images of women, Mary De Morgan's hybrid figures, the Rainmaiden or the Windfairies, are highly mobile and transformative, raining, dancing, flooding, vaporizing, and blowing at will. By the end of *The Windfairies*, De Morgan has constructed an ecology that fosters respect for the elements and for a vision of the New Woman artist. As the ethos of Arts and Crafts and the practices of New Woman literature and culture converge, De Morgan's tales link women and the elements materially to prove neither entity is a mere resource. Woven into narratives that resist the commodification of women's arts or even of plants in gardens, the elements move the tales toward the "sexual anarchy" of a more equitable society that fosters all things' capacity to act, generatively and creatively.[75] De

Morgan makes way for a feminist, nonreproductive immanence, exploring a different type of generative power: women's artistry and creativity alongside elemental vitality. Finally, the elemental figures in De Morgan's tales make it clear that while they may aid humans, they do not exist merely to serve them. As vibrant "actants," they are capable of harm as well as help.[76] One of the text's most timely ecological insights is that to forget the power of the elements, as humans often do in the tales, is to risk harm and perhaps to visit "great misfortune" on what we "love best in this world."[77]

Cloud

De Morgan's tale "The Pool and the Tree" explores the elemental link between plants and water. The story begins with the loving symbiosis between a tree and the pool of water that feeds it in "the middle of a vast wilderness."[78] "Beneath the shade of its branches was a little pool, over which they bent." The tree and the pool are genderless partners and "beloved friends." "But for you and the shade you give me I should have been dried up by the sun long ago," says the pool. "And if it were not for you and your shining face, I should never know what my boughs and blossoms were like," answered the tree, aware that without its water source "I shall die."[79] For years, the pool and the tree freeze, "die away," and return to life each spring.[80] Soon, however, "there rode over the moorland a couple of travelers in search of rare plants and flowers."[81] Instrumentalizing the landscape as their hunting ground, the men feel cheated. They take advantage of the tree's shade in order to rest and complain about the landscape, which offers nothing "very rare."[82] Concerned only with their own physical discomfort, they grouse, "'We have not found much,' . . . it seemed scarcely worthwhile to come so far for so little."[83] The other collector notes the difficulty of their endeavor: "One may hunt for many years before one finds anything."[84] Just as he says this, however, his partner "picks up one of the fallen leaves of the tree which lay beside him, and at once he sprang to his feet and pulled down one of the branches to examine it." His partner "also closely examined the leaves and blossoms . . . declared this was the best thing they had found in all their travels."[85] Soon the men leave and return with shovels and pickaxes. They "wrench" the tree out of the ground and away from the pool to bring it to a "great garden" on the other side of the world. In the garden, the tree is transplanted, fenced off, and tended by a gardener whose sole task is to care for it. Devastated at the loss of its friend and keenly aware that the removal of the tree will mean its own evaporation, the pool transforms itself into a cloud and travels the earth searching for the tree. In its wake, the pool leaves only a "dry hole."[86] Finally, the two are reunited as the pool blows over the garden and rains on the tree in the night, its water pooling around its roots. There the pool remains hidden and never appears above the ground again. Delighted at the tree's recovery, the gardeners don't know that its water source has returned. The tree and the pool remain secretly reunited. They become a popular exhibit in their garden, important because the tree is a great "rarity" in the plant world.[87]

In this, surely one of the earliest critiques of imperialist plant collection, De Morgan digs deeply into the hidden systemic violence of the practice, showing how it destabilizes a "beloved" symbiosis of tree and water, which, in turn, throws a former oasis into deforestation and drought. As she transforms the tree's water source into a cloud that seeks out its lost friend, De Morgan represents the plant and water becoming themselves through a process of mutual aid and transformative encounter. In this way she connects the sky to earth and the plant to both. In contrast to representations of cloud and sky in other Arts and Crafts works—those of Edward Burne-Jones, for example, to whose grandchildren Mary De Morgan told and dedicated this collection of tales—her ecological vision becomes even clearer. In Burne-Jones's work, representations of clouds are used to create depth, gravitas, and light for the human story in the painting. Often shrinking the sky to a small rectangle, the better to focus on the earthbound quality of the human, mostly female, figures in his paintings, Burne-Jones's skies create a very stable perspective, parallel to the plane of the painting. The sky above creates a low perspective as the viewer looks up to it in paintings such as *The Golden Stair*. Dark skies might signal that "the storm-cloud of modernity" had arrived and with it anthropogenic climate change.[88] Nonetheless, such images of clouds in literature and art remain, as they usually do today, "framed as if encountered by a human viewer."[89]

De Morgan's tale plays against that standard frame. Pushing the human story to the margins, the tale's horizon where earth and cloud meet is permeable and unstable; cloud moves from sky to ground below, to plants and "damp" surface above. Animating the clouds and making them the central protagonists, this is a nonbinary love story in which the cloud desires the tree's roots and they reciprocate, desiring the cloud. The tree is "what I long for," cries the cloud. "For then I can lie in the dark where no one may see me, but I shall be close to my tree, and I can touch its roots and feed them, and when the raindrops fall from its branches they will run down to me."[90] De Morgan's vision of the material world recalls medieval tales of cloud ships or "the sea above."[91] Such a tale "does not simply reaffirm human primacy but conveys" the reader's eye away from the standard stable horizon line, where earth meets sky and separate ontological categories stabilize the view.[92] "The Pool and the Tree" moves the mind's eye away from the horizon and toward a more exuberant, elemental ecology. Even as the tree is examined as a "thing" by the collectors, the relationality and sentience of the pool and the tree's ecological partnership is affirmed by the constant up/down movement of the story: "The pool lay looking lovingly up at the tree, and the tree gazed down at the clear water of the pool, and they wanted nothing more."[93]

The symbiotic partnership between both is further developed in Olive Cockerell's full-page illustration of such moments, which deploys both the strong lines of Arts and Crafts and the fluid influences of Art Nouveau (see fig. 4). In one image the tree and pool blend into each other. The tree figure's hands and feet disappear into the leaves and limbs of the tree itself. The roots of the tree flow into the pool's hair and both melt into the water, becoming the ripples of the pool. Both figures are nude, and while the tree figure is represented looking down toward the pool, the water is represented as stretching up toward him, placing her hands upward around the trunk of the tree. Likewise, the tree, with its canopy open to the sky, signifying the "lacy" airiness De Morgan describes, is permeable to the elements of air and water that flow through and into both. Both figures are vulnerable in their nakedness, stripped of any barrier to their embrace. Their facial expressions register profound tenderness and the melting of their bodies into each other signifies the "cross-ontological alliances by which eco-systems thrive, change, commingle, create."[94] In looking at the illustration, one cannot see where the tree and the water begin and end; the intimacy of their entanglement is expressed both in De Morgan's simple language and Cockerell's image, the alterity of which prevents both from descending into sentimentality. Cockerell's illustration, like De Morgan's tale, represents the tree and pool as two beings worthy of a love story; their tale is, importantly, not allegorical, but traces the elements as vibrant beings worthy of their own story.

However, in the tale the two are never gendered; each is always referred to as an "it." In seeking the tree after it is taken, the cloud informs others, "I cannot stay, I am seeking a lost friend . . . and it scudded past them, leaving them to roll over and over, and tumble about, and change their shapes, and divide and separate, and play a thousand pranks."[95] These "its" are a clear example of an exuberant, generative, nonheteronormative Nature, and as such they may represent what queer ecologists like Stacy Alaimo describe as the "exuberant pleasures of thinking with, and feeling with, an abundantly, uncontainably queer world."[96] As the tale shifts the aesthetic limits of cloud and sky, so

Figure 4. "The Pool and the Tree," Olive Cockerell, from Mary De Morgan's *The Windfairies* (1900). (Courtesy of Taylor Bowman, Auburn University)

does cloud cover what Sandilands terms "discursive frameworks" of desire established by heteronormative culture.[97] The materiality of things in the tale refuses to fit into a heteronormative plot. The two beings remain, particularly in contrast to the plant hunters, more than human and without recognizable genders. That framework is burst by love and friendship. And as the reader (or listener) is encouraged to bond with "our cloud" against the terrible plant hunters, invited to see what the expert gardeners at the "great garden" cannot, the narrator invites the reader to fall in love with the exuberance of the more-than-human world as it scuds, rolls, tumbles about, changes shape, divides and separates, and "plays a thousand pranks" on our expectations.

Throughout the tale, then, the reader looks up and down, where the action is, in the air, on and under the earth. It is terrible to witness the removal of the tree as the pool "for the first time" looks "straight up at the sky without seeing the delicate tracery made by the leaves and twigs against the blue, and it called out to all things near it," mourning and lamenting. "Poor pool," consoles a swallow, who tells it to learn from the moon, sun, and clouds how to transform itself. At the cloud's request, the sun "threw down hundreds of tiny golden threads," which change the pool until it grows "thinner and lighter," "rising through the air, till at last it had quite left the earth, and where it had lain before, there was nothing but a dry hole, but the pool itself was transformed into a tiny cloud, and was sailing above in the blue sky in the sunshine."[98] The air is now full and consequential rather than ethereal. Inhabited by birds, "our cloud," and planets, sky and earth cease to be "mutually exclusive domains of habitation."[99] Each of these agentic beings seem threatened as ecological awareness is formed. As the tale is written, moreover, one would need to be more than jaded not to sympathize with the elemental beings and rally against the plant hunters and gardeners. Ironically, "natural history" and knowledge itself—traditionally represented by such botanical experts and their collections—is rearranged along with the horizon.

At the end of the tale, then, the reader is given eyes to see what the latter experts can't, a commingling, erotic and graceful: "The wind blew the cloud lower and lower, till it almost touched the top branches of the tree. Then it broke and fell in a shower, and crept down through the earth to its roots, and when it felt its drops the tree lifted up its leaves and rejoiced, for it knew that the pool it had loved so had followed it." "Have you come at last?" asks the tree. "Then we need never be parted again."[100] The listener has now become the naturalist and knows better when De Morgan describes the gardener's reaction to the tree's recovery: "In the morning when the gardeners came they found the tree looking quite fresh and well and its leaves quite green and crisp. 'The cool wind last night revived it,' they said, 'and it looks as if it had rained in the night, for round here the earth is quite damp.' But they did not know that under the earth at the tree's roots lay the pool, and that was what had saved the tree." The reader's new ecological consciousness of the tree and the pool is brought into the present and even the future through the tale's ending: "And there it lies to this day, hidden away in the darkness where no one can see it, but the tree feels it with its roots, and blooms in splendor, and people come from far and near to admire it."[101] Those spectators, however, view a rootless modern botanical drawing; they can't see below the epistemological frame, to the roots. This remains a secret vision now only shared by a newly linked ecological community—the tree, the pool, the narrator, and the reader.

Investing the tree and the pool with abiotic integrity through narrative, the tale accomplishes what science narratives or socialist tracts cannot. It informs, but it also materializes the abstraction of Nature as thick, agentic beings. The tale clearly critiques the linking of natural history and imperialism, collection and acquisitiveness, the idea that plants are collectible without consequences. The dark hole left behind when the tree and pool are gone, having been dug up and transplanted to the "other side of the world," aptly makes what Rob Nixon terms slow violence, the impact of deforestation on colonized lands, visible.[102] Nixon describes slow violence as a "violence that occurs gradually and out of sight, a violence of delayed destruction that is dispersed across time and space, an attritional violence that is typically not viewed as violence at all."[103] De Morgan's gentle and melancholy tale and its bittersweet ending represent what Rob Nixon describes as a scarce presence in Western art, the dramatization of slow, attritional ecological exploitation in which all of the damage is not immediately apparent, but remains hidden to the European observer. Moreover, the very standard of the tree's collection as a "rarity" underscores the trouble with colonial plant collection, which simply does not value ecosystems but only the "rare" individuals that will complete a collection at a place that looks much like Kew Gardens. De Morgan's tree and pool, in contrast, are represented as of great value not in isolation, but in mutual interaction with each other.

De Morgan's vibrant elemental narrative acts with her critique of botanical imports and "specimen" gardening to contest the dominant garden and botanical epistemology of her time, in which the appearance and "rarity" of plants rather than their ecology are the currency of knowledge. By default, the tale underscores early green appreciation for native plants and the integrity of local environments. Animating ecological knowledge, De Morgan critiques botanical epistemologies as acts of both local and global violence. The ecocritical legibility of her tale asserts that no practice is "beyond the call of environmental readings. . . . Matters of environmental concern and wonder are always 'here,' as well as 'there,' simultaneously local and global, personal and political, practical and philosophical."[104] As women are traditionally associated with both water and trees in Western culture—and, therefore, as Greta Gaard argues, are all three more easily instrumentalized as resources—De Morgan's nongendering respects both the elements and the tree's alterity from humanity.[105] By refusing to ascribe "natural" categories of gender and sexuality to the nonhuman natural world, she constructs a queer ecology that cuts Nature loose from conventional reproductive sex/gender roles. Instead of affirming the exploitation of the tree and the pool through ascribing gender roles, De Morgan thickens their entanglement and relationality on their own terms as alternate living beings.

At the time "The Pool and the Tree" was written, plant and "specimen" collection still formed the basis of knowledge in the discourse of natural history, which was slowly being replaced by the new science of ecology and an understanding of ecosystems. The world of exotic plants still "beckoned. Botanists traveled abroad to compile information about plants in British dominions, and collectors and colonists sent plants from Canada, India, and elsewhere, to be stockpiled at Kew, the botanical heart of the Empire."[106] While such displays and practices "gave women areas for growth and strategies for observing, learning, teaching, and writing about nature," the imperialist and antiecological elements of plant collection were environmentally devastating.[107] As residents of other ecosystems, such plants were highly vulnerable to disease and required, like Blanchelys's rose, constant gardening, often at great expense. De Morgan's tale, however, critiques the *ecological* cost of the specimen garden, representing the cost of England's collection practices upon ecosystems and the new ecological concept of life as a "web" or a mesh. "The Pool and the Tree" focuses on abiotic environmental factors and relationships: ecological cooperation, instability and equilibrium, competition, cooperation, climate, integration, individuality, and "community," the precursor to the contemporary term "ecosystem."[108] The tale enchants the science of ecology, writing the cooperation of plants and the elements as a love story through the compressed form of the fairy tale, while the Victorian specimen garden is remythed as a site of hidden violence.

De Morgan's critique of the specimen garden reveals that the practice fundamentally violates the symbiotic relationship between living things, the elements, and climate. The transformative and mobile element of water is entirely underestimated by the humans in the tale. This becomes a theme in *The Windfairies*, in which humans not initiated into the powers of the elements fail to appreciate their material power, their ability to travel great distances, establish equilibrium or instability in the environment, and facilitate the integration of individuals into ecosystems, and in doing so or not doing so, help or harm them. Reading De Morgan's tales, one becomes intimate with these ecologies. The reader becomes complicit with the woman storyteller, placed in a superior position to the male expert naturalists and gardeners, a trend that continues in other tales with authority figures like kings and queens. Instead, the reader is allied with the pool, the tree, and "Mrs. Oakchest," the pseudonym Mary De Morgan used when telling tales to Angela, Dennis, and Clare, Edward and Georgiana Burne-Jones's grandchildren.[109] The conspiratorial nature of the tales is signified by the dedication on the book's frontispiece as the text rises up from Cockerell's drawing of the chest inscribed with "MRS" on the inside and "MDM" on the exterior. A key with a heart-shaped bow floats next to it, mid-page, indicating that the heart unlocks the key to "these little tales." By establishing a new form of knowledge as well as empathy for other beings, De Morgan's tales open her readers' minds wide to the elements. As the key floats free next to the chest, one cannot help but think of the New Woman's iconic latchkey that opens Pandora's box, providing her autonomy and free passage as well as access to a whole host of new natures.

De Morgan's ecological magic aligns her authority as a woman writer with the new natures of the New Woman and the early green impulses of the culture of Arts and Crafts even as she critiques that movement's favorite ecological paradigm of the garden and the limits of botanical knowledge then considered an important achievement for "ladies." Her tales link women and the elements more intimately as *The Windfairies* progresses. Trading the mobile elements, a Rain Maiden and windfairies for the static Floras and Pomonas so typical of Arts and Crafts makes De Morgan's tales increasingly nonreproductive, imbuing them with modern qualities of speed and mobility. As Mary De Morgan leads the reader out of enchanted gardens and into tales of women enchanted by the elements, she develops a fresh understanding of immanence that is strikingly free of the biological determinism associated with the garden. Instead, her human characters develop an awareness of the extent to which they merge with and are accountable to an elemental world that is both inside them and outside them, a place apart and within. Ethical choices in De Morgan's last collection of tales are shaped by encounters with a multitude of bodies, human, rain, cloud, and air.[110]

Rain

The narrator's perspective in "The Pool and the Tree" makes a case for the first principle of conservation biology—that is, while the tale advocates for the idea that "there is intrinsic value" in "the complexity of ecological systems," it also makes a case for relational value.[111] The specimen garden in the tale violates both and the foundational land ethic Aldo Leopold would voice in 1949: "A thing is right when it tends to preserve the integrity, stability and beauty of the biotic community. It is wrong when it tends otherwise."[112] "The Rain Maiden" critiques the failure of humans to understand either intrinsic or relational value when they forget about the rain. In De Morgan's tale, the elements are an alternative form of nonreproductive, generative energy and desire. But in this tale, women's creative integrity and elemental agency are allied. The Rain Maiden wants only to dance and declares, "I shall never be the wife of any man."[113] Both a daughter and an element, she experiences the pleasures of immanence as a "maiden." By the end, De Morgan's tale indicates that it will not pay to forget promises made to the rain or to daughters.

The tale begins during a great gale when "the raindrops sounded like the knock of a hand that was knocking to gain admittance" on a cottage door. The woman inside opens the door and greets a "tall woman wrapped in a grey cloak with long hair falling down her back." After offering the visitor a glass of water, the cottager is asked what she would like in return for her kindness, because, although the grey woman comes "from quite near," many would not have admitted her.[114] Soon the cottager begins to stare at her visitor, for she realizes her apparently "dull grey cloak" is "covered with jewelry,— clear stones, like diamonds with many flashing colors."[115] She asks for only one thing, a child. She would care for the child and love it, and then, "in turn . . . it would love and care for me." "And if you had a little child," the grey woman reflects, "you think you would really love it better than anything in the world. Many women say that, but few do it." Despite her doubts, the grey woman grants the cottager's wish. But, she warns, when the cottager loves "anything else better than your little daughter and her happiness, it will go from you; so remember my words."[116] As the grey woman mixes the pronouns of the future child from "she" to "it," so is the baby that is duly born a hybrid figure. On the night of her birth, "the wind howled, and the rain fell as fiercely as on the night when the grey woman" had appeared. And, as soon as she can, the child dances in the rain and worries her mother who cannot "cure" her of this. When the girl hears "the rain pattering against the window-panes she would cry, 'Listen, mother, listen to my brothers and sisters dancing,' and then she would begin to dance too."[117] Soon, the girl attracts the love of a prince and her mother becomes "nearly wild with joy."[118] But the girl does not want to marry. Even as her mother becomes angry and accuses her of madness, the girl declines. Ignoring her refusals, a wedding date is set, and the rain begins to fall hard as it approaches. When the river swells and floods its banks, "everyone was frightened, save indeed the shepherd's daughter, who went out into the wet and danced as was her wont, letting the torrents fall upon her head and shoulders."[119] The bride steps through pools of water on her way to church. Once the ceremony is complete, a great hurricane floods the palace and the bride disappears. But long after the prince has married another, he can hear her voice when it rains.

"The Rain Maiden" reflects upon the materiality of rain and human attempts to relegate it, and women's desires, to nothing of intrinsic value. As rain drives narrative through its mobile, fluid, and agentic power, water and women ally and speak back, laugh, in fact, at the human inability to understand their own vulnerability to rain and their intimacy with it. This is clear to the girl who is alternately a "she" and an "it," erasing through her being the difference between flesh and water. The people in the tale who fear or deny this elemental intimacy are also fearful of the rain. When the cottager insists upon upholding the logic of the marriage market and forgets to follow through on the promise she has made to the grey woman, she forgets the power of rain over people. By contrast, her daughter is tapped into it. In the elemental tradition of Arts and Crafts, she engages the rain as an artist, a dancer. Her plot moves her forward to that ecstatic intimacy rather than to marriage.

The Rain Maiden's insistence upon her sexual integrity and bodily autonomy is communicated through speech and dance. In Cockerell's illustrations, and in De Morgan's text, her movements resemble expressive dancing in the style of Isadora Duncan (see fig. 5). She is pictured barefoot, in a loose tunic, and uncorseted, dancing solo in the rain, lifting her hands to a sun shower, or dancing upward and away from a suitor. Water flows over her feet, and her hair flows like the water of the river. This solitary dance divorces dance from its heteronormative partnered history and its link to the practice of courtship. Like Duncan, who performed in Europe the year *The Windfairies* was published, she too seems "The dancer of the future," who "shall dance the freedom of women."[120] Such fin de siècle dancers capture "the anxiety over the New Woman and her threat to male subjectivity," as her rejection of partnered dancing was "modern" and tantamount to "her liberated expression of sexuality and need to engage in the public sphere . . . as a threat to traditional gender roles and

Figure 5. "The Rain Maiden," Olive Cockerell, from Mary De Morgan's *The Windfairies*, (1900). (Courtesy of Taylor Bowman, Auburn University)

thus to the larger social order."[121] Dancing alone, for her pleasure, in the rain, the Rain Maiden refuses to commodify her dance as a courtship instrument. Communing with her "brothers and sisters," the raindrops, she eradicates the distance between culture and nature and performs a remarkable reversal of the traditional fairy-tale plot. This tale does not end in marriage but in another elemental transformation as the girl transforms into rain itself.

In text and image, the tale's intimacy with the rain muddles the horizon, the directions of up and down. Not only does the tale aestheticize an often unwelcome elemental form, but it bursts limits and oppositions of all sorts. De Morgan defies the very opposition between "rain and shine," when "the sun came out while the rain was falling."[122] Dancing in a sun shower, the girl seems like the grey woman, "covered with jewels of every color, clear and bright."[123] Even so, she desires to remain where and who she is: "I will never live at the palace, and I will never be a Queen."[124] As she will not serve her mother's social ascendance, she critiques the idea that water is a mere resource for people. When the prince, a "grand young man dressed all in velvet and gold," first arrives on the scene in a boat to fish, the girl questions who he is and what he is doing on the water, "for she was afraid of no one. . . . 'It is cruel to take fishes out of the water, leave them alone and come and dance on the bank with me.'" Guarding her autonomy, she tells both her mother and the prince, "I will never be the wife of any man."[125] But the mother will not listen until a hurricane reminds her of the power of water.

In the literary fairy tale, the plot of a young girl's life is often predictable. The reader empathizes with impoverished heroines, like the Little Match Girl, and expects a story to end with death or the arrival of a prince, followed by a wedding. But rain very often surprises people. Lowell Duckert argues that "rain washes away the illusions of human centrality and analytical enclosure."[126] Even as the rain pelts the roof of the church during the ceremony, it is a great shock to the people when the hurricane comes. As the ceremony begins, the wedding guests ignore the rain on the roof trying to drown out the voice of the priest, "and when they went into the castle to the banquet, the water burst through the doors opened to receive them, so that the King and the wedding guests had hard ado to keep dry."[127] Nonetheless, they continue with the "grand feast" until "just when the goblets were filled with wine," the flood comes. Suddenly all human plans evaporate, as "in one moment, the rooms were filled with water, and no one thought of anything but to save themselves." When the hurricane subsides, the bride is gone. Only then does the bride's mother remember the words of her elemental visitor: "When you love aught on earth better than your daughter and her happiness, she will go from you."[128] Although people will themselves to forget the power of rain, its appearance in magical and material form is a constant reminder that their transcendence of the weather is a fiction. The human mother who receives a magical gift, a child, in exchange for giving a simple glass of water, forgets where water comes from and forgets that she owes it a debt.

De Morgan's "Rain Maiden" crosses boundaries between watery and human bodies. The heroine seems a hybrid, a New Woman, more-than-human, evolved well beyond her mother. De Morgan's representation of the power of water both naturalizes the shock of social change itself—the birth of a girl who refuses to marry—and overturns the idea of a merely reproductive Nature. While water is typically associated with women and maternity, it is quite clear here that De Morgan wants none of that. In their failure to imagine other natures, the human characters in the tale must finally stop marriage plotting and save themselves. They are about to get wet. Almost like an anarchist's bomb in a dynamite narrative, the hurricane scene disrupts the public space of the royal hall and the private space of marriage, motherhood, and domesticity. The beating of the rain suggests that something elemental objects to the marriage plot as . . . unnatural.

In the slippage between rain, grey lady, and Rain Maiden, there is an opportunity to remember the agency of both rain and women. Jane Bennett asks, "Does life only make sense as one side of a life-matter binary, or is there such a thing as . . . a life of the it in 'it rains'?"[129] The wishes of the rain and the maiden parallel each other in the story, sharing intent and purpose, the definition of agency. Bennett writes that such a "touch of anthropomorphism . . . can catalyze a sensibility that finds a world filled not with ontologically distinct categories of beings (subjects and objects) but with variously composed materialities that form confederations" and reveal "similarities across categorical divides . . . parallels between material forms in 'nature' and those in 'culture.'"[130] At the end of the tale, the readers finally accept the parallel agency of the girl and the rain because the two speak together: "Come and dance with us, and feel our drops upon your face."[131]

Wind

Mary De Morgan, like us, lived in a time that failed to account for the value of rain or air. Then, the sky might carry waste with impunity. Now, water and wind may be instrumentalized as in De Morgan's day, to power mills and run factories. Then as now, water and air were "free," "resources" to be used at will. The elements still have no place in the economy, no value in and of themselves, and, as Iovino and Oppermann argue, they do not "'figure in environmental ethics or politics," although environmentalists and ecocritics hope they will "in the near future."[132] When water or air "rights" are discussed, we mean only the human right for access to them. Our vulnerability to water and wind, moreover, is consistently underappreciated. In "The Windfairies," De Morgan addresses the invisibility of the agency of air, its power, and human vulnerability to it. Because the wind is omnipresent and necessary to life, there is no "outside" or "away" from it. No one sees the windfairies, however, in the tale, except the child Lucilla.

As Lucilla watches her father's windmill, she alone sees what fills its sails: airy "grey elves" who dance exquisitely. Fascinated by the windfairies' airy dance, Lucilla begs that they teach her to move like them. At first the windfairies are wary. Gray elemental creatures, the stepchildren of a thermodynamic age, they do not think much of humans. Striking the same slightly exasperated tone as the grey lady in "The Rain Maiden," they explain their elemental power to the human child: "We blow the winds and sweep the earth. When there are many of us together we make a great hurricane, and human beings are frightened. We it is who turn your mill wheel for you and make all the little waves on the sea."[133] When she insists that they can trust her, they reply "that is what all mortals say." Because she seems talented and brave, however, the windfairies relent. In accepting their gift of dance, Lucilla is entering into a bargain with potentially dire consequences. If she ever tells "any mortal who it was that taught you how to dance, you will never dance again, for your feet will become heavy as lead." Additionally, some "great misfortune will overtake whatever you love best in this world." On the other hand, if Lucilla "keeps faith" with the windfairies, they "will never forget you, but will come to your help in your direst hour of need."[134] In making this bargain, the wind fairies and Lucilla create an alliance that is reciprocal and respectful, mutually beneficial, but uneven. The wind holds elemental power over Lucilla.

Lucilla grows up to be a rare dancer who times her movements to the waves and breezes. But after she chooses a husband, the best dancer of all her suitors, Lucilla falls upon hard times. He must go to sea and she is left to care for her children and father. One day, while dancing on the beach, she is approached by two men and offered a chance to leave her home and children to travel to another kingdom to perform for a King and Queen. She agrees to go and bids goodbye to both her children and the windfairies. Lucilla knows the risks she will be taking and she goes out "alone on to the cliffs, and stretches[s] out her arms." She calls to "to the windfairies to go with her and help her, for she feared what she was going to do, and she begged them to be true to her, as she had been true to them."[135] The windfairies do not answer, but the ship is blown to the new kingdom where her loyalty to the windfairies and her own artistic integrity will be tested.

Once in the new country, Lucilla finds that the royal couple want to dress her in "gold and jewels." She responds by asking to wear "a light grey gown like the windfairies."[136] Before she dances, Lucilla evokes the windfairies' help: she "threw wide her window and held out her arms, and cried out, 'Now help me, dear windfairies, as you have done before; keep faith with me, as I have kept faith with you.'" Once again, the fairies do not answer but Lucilla dances brilliantly. She commands public space alone, accessing the element of air within and without her, and translating her innermost thoughts and emotions into movement: ""Light as the sea foam . . . when she swayed and curved to the sound of the music, it seemed to her as if she heard only the swish of the waves as they beat upon the shore, and the murmur of the wind as it played with the water, and she thought of her husband out at sea, with the wind blowing his ship along, and of her little babies living in the cottage on the beach."[137] When Lucilla finishes dancing, "the musicians one and all lay down their instruments, and rose together, clapping and applauding."[138] Lucilla has partnered with the wind.

Soon the Queen grows jealous of Lucilla's dancing and demands that she tell her the secret to her skill. Lucilla refuses. The Queen insists; she wants to be "just like Lucilla."[139] When Lucilla explains to the king that she learned by watching the windmill, he is contemptuous: "'Nonsense,' no one could learn dancing from looking at windmill sails, neither was it possible that she, a poor miller's daughter, could have learnt such dancing by nature."[140] Arguing their right to both collect and imitate Lucilla, the King and Queen miss the source of her skill in her elemental liaison with her teachers and mentors. It was Lucilla's ability to see the air and converse with it as a more-than-human equal that made her an artist.

As neither Lucilla nor the air can be bought, the royal couple are displeased and they accuse her of witchcraft. The showdown that ensues tests Lucilla's loyalty to the windfairies. On pain of death she is ordered to tell the secret of her skill. She is offered a bag of gold so heavy she can hardly lift it. She is imprisoned and tortured. When the King and Queen try to drown her, however, the water "held her up; and the waves rocked her gently" back to land. Sentenced to be burned at the stake, Lucilla stands on the pyre but "the flames divided on each side, and were blown away from her all round, so she sat in the midst quite unhurt." Olive Cockerell illustrates this scene in the marketplace and visualizes the presence of the windfairies invisible to both Lucilla and the crowd. They circle around her, touching her hair protectively and pushing the flames to each side of her chained body. As they caress her, Lucilla bows her head and clasps her hands in front of her in supplication to the wind. Even so, Lucilla's body is robust and strong. The windfairies blow swirls of smoke away from Lucilla, while her hair and tunic swirl into and through the drapes and long streaming hair of the windfairies. Even as Lucilla is bound and stands still, Cockerell's representation of the wind expresses her mobility; the image itself appears almost to burst out of the frame in the center of the page.[141]

When the fire fails, Lucilla is given one more chance to sell the secret of her art. If she refuses now, she is to be beheaded. "A large bag of gold is placed on the scaffold," and the King offers "that if she would confess, even when she was upon the scaffold" she will be freed. Soon the wind arrives, and Lucilla hears it "roaring, and the trees were bending and breaking in the gale." As she is led to the scaffold the guards "had much ado to get along, for the wind blew so hard that they could scarce keep upright in it." "Now dear windfairies," says Lucilla, "help me for the last time, and keep faith with me, as I have kept faith with you." As "the executioner lifted the axe in the air," there "came a sudden roar of wind, and the axe was swept from his hand, and the houses in the market-place tottered and fell, and high up on the hill the palace was a mass of ruins."[142] At the conclusion of the tale, Lucilla retains her skill, her family, and her artistic integrity. She also brings home the bag of gold.

As the windfairies' protégé, Lucilla is sworn to her alliance with them, even when she cannot see them. In return, Lucilla enjoys their gift, but she also experiences the same risks that they do—that she will be instrumentalized, co-opted, or commodified. Her gift brings her ecstasy and provides the very quality of originality and artistry that attracts the envious eyes of others, "to those who watched, it seemed as if she and" the wind "were one together."[143] Lucilla dances for self-expression, not to invoke the Victorian norm of the ballroom with its "strict rules of etiquette" and its practice of partnership which aestheticized the marriage market and courtship.[144] Her dance taps into a historic moment when women's solo dancing, such as Isadora Duncan's, could "facilitate free expressions of sexuality for the dancer as well as promote artistic experimentation."[145] Lucilla explores what it is to be an artist and so De Morgan writes a new plot for the Victorian fairy tale. Although De Morgan is often compared to Hans Christian Andersen, she does not punish her dancer for her gift as in his "The Red Shoes." Her marriage does not end her story, nor does she sacrifice her artistic integrity even to be reunited with her children. Lucilla's beauty, moreover, resides in her skill, her artistry, in what she does. Less powerful than the wind, she seeks only to be its loyal partner. The fairies, wary, less than accessible although she calls to them often, maintain their elemental presence as more-than-human beings. The windfairies adorn the cover of De Morgan's book, towering over the free-floating female body of a human woman who holds a tiny windmill in her hand. The windfairies make their power clear; these are elemental partners that matter.

Lucilla clearly embodies "the dancer of the future" that Isadora Duncan says "will be one whose body and soul have grown so harmoniously together that the natural language of the soul will have become the movement of the body."[146] Her dance separates "the dance, a sensuous, physical activity from the ball" and the marriage market.[147] As Lucilla embodies wind and water, she displays the entanglement of nature and culture rather than their division. As a dancer, she does not defy gravity, but expresses the agency of air. Endowed with the windfairies' gift, Lucilla dances to valorize their creative partnership. In the tale, the elements enter their human subject, who becomes, for all intents and purposes, posthuman, in that she is coterminous with rather than distinct from the environment or elemental energy. De Morgan's elemental narrative reminds the reader that human beings are never without the elements, but always in their company. In the company of the elements and containing them within her, Lucilla learns that they will be true to her if she refuses to harness them for profit. Her tale of the artist's creative alliance with the elements reveals her intimate knowledge of the pains and pleasures of making art, particularly for the woman artist, as she negotiates those who want merely to copy or reproduce her creativity without fully engaging the will and power of its source. The wind is not Lucilla's to sell. Trusting alliances between women as her subject, De Morgan makes way for other kinds of plot and other kinds of love.

By reviving an elemental ecology in the midst of industrialization, which considers itself to have made the elements obsolete or a handmaiden to industry, De Morgan expands "the limits of anthropocentricity and the intimacy of narrative-making" to ecological ethics.[148] She could be charged with anthropocentrism, because she focuses on the artist's crisis and rescue. Or the fairies could be considered anthropomorphic in their appearance and in their dancing. However, in turn, the tales foster relationality and intimacy between human and elemental creativity. Certainly, De Morgan's elemental ecologies in *The Windfairies* foster a kind of humility in which humans realize that they are not the sole owners of agency. The archaic epistemology disrupts human superiority and resists the anthropocentric view of either women or Nature as a mere resource.[149] As De Morgan's tales place human and elemental figures in interdependent relationships with each other, giving voice to both the elemental and the human, the intrinsic and relational value of both is affirmed. The elements become not allegorical but vibrant things, receiving stories of their own.[150] Their plots create a body of Nature that accounts for their power.

This gendered elemental shift reminds us of how much more there is to learn about New Woman literature and culture. In 1979, Anthea Callen called for further attention to the women artists of the Arts and Crafts movement. However, when Jan Marsh revisited the issue in 2002, she argued that the field remained still unexplored.[151] Only in 2017 was an exhibit and catalog launched on the most well-known of Arts and Crafts women, May Morris.[152] De Morgan's tales are, then, an especially timely and welcome gift for scholars today. As she critiques and replaces the garden natures she inherited from the Arts and Crafts Movement in exchange for startlingly new transformative, elemental ecologies, De Morgan's voice speaks to our time as well. She boldly redefines what Nature is and what counts as the environment, expanding the concept to include far more than any one particular, sacralized place apart. In the tradition of the New Woman, she redefines Nature itself, claiming it as less a place than a process, one that fosters women's creativity rather than their maternity or service. In "The Windfairies" and "The Rainmaiden," women, water, and wind share the experience of diversion and the definition of vocation as commodification. When faced with the narrowing of their mobility, they ask, "Why should you stop me?"[153] This question is taken to another level in the next chapter, when *The Windfairies'* illustrator, Olive Cockerell, changes fields and asks the same question of her famous aunt, Octavia Hill.

2

"WE ARE TWO WOMEN"

Alternative Agriculture, the New Woman, and Rural Modernity

While Mary De Morgan rewrote the garden in her tales, the second generation of Arts and Crafts women sometimes re-created it from the ground up. They tested the idea that everyone should do their "fair share of manual labor, that the workers should be thinking and that the thinkers should be working."[1] This principle converged with a rise in horticultural education for women. Demanding, for example, that no profession for women should be considered "taboo" or "unladylike," Lady Frances Warwick argued that "the digging of a trench or the earthing-up of the plants" should "not be limited to one sex."[2] Women like Olive Cockerell likely came to agriculture through this path, championing the inherent dignity of manual labor and spurred on by new opportunities provided by a wave of new training programs for women that began near the end of the nineteenth century.[3] These programs, viewed suspiciously by Octavia Hill, sought to combine the "New Woman and Old Acres," to the advantage of both.[4] Growing vegetables or fruit, women might pursue a sustainable, independent, "health-giving" life.[5] They might reasonably profit from a hunger for fresh, local food during an age of mass importation.[6] While the dimensions of market gardens were small, the conceptual and material shifts they promised were grand. For Cockerell and Nussey, their garden meant that they could experiment with a new profession for women, to see if "two women like ourselves" could "make a living in a small holding."[7] After a bicycle tour scouting locations, they settled on a few acres in Sussex, both "so near the hum of civilization" and "quite hidden in the woods, entirely to our liking, for the life we were going to lead was to be free."[8] From 1907 to 1910, in their small holding they pursued that goal, constructing a green future for women through the practice of alternative agriculture.

The latter is so called because it sought to provide an alternative to the ecological and economic failure of monoculture in England during the long agricultural depression that arced over the period of the New Woman from 1879 to 1939.⁹ During this time, English corn, which had long displaced local fruits and vegetables, was undersold by cheap imports from Canada, the United States, Australia, and New Zealand. Having widely neglected their local orchards and vegetable gardens to concentrate on corn themselves, the English now became dependent on importation for grain, fruits, herbs, and vegetables. Their farms were both too large and alternative crops too unprofitable to allow them to rethink practices and acreages formed during the Age of Improvement. The cheapest solution was to turn arable land to pasture. In the meantime, greengrocers concentrated on creating a "Goblin Market" of imported produce, featuring "more expensive and exotic vegetables like asparagus" that could fetch a good price.¹⁰ Soon basic fruits and vegetables fell out of working people's diets; thousands of farmers emigrated or chose other professions.¹¹ To address the absence of local fresh plant food between 1870 and 1919, working men's allotment gardens were created, and in 1904 a delegation of Evesham gardeners traveled to Paris to observe the French market gardening system.¹² They noted the French practice of companion planting several varieties of vegetables in the same bed and that the French "make use of every inch of room. . . . Instead of getting one crop out of the beds they get four."¹³ C. D. McKay, author of *The French Garden: A Diary and Manual of Intensive Cultivation* (1908), proclaimed that "French gardening" in a city like London might be the means to "supply ourselves" with food.¹⁴ Reading "old English books on the subject," McKay noted that the British once grew "early vegetables" and could do so again; he deduced that "as soon as the British people realize that they can have fresh salads and '*primeurs*' grown in their neighborhood," there will be a "large demand."¹⁵ If currently such produce was considered only in the reach of the affluent, it was "within the reach of the bulk of families."¹⁶ McKay's handbook, the first practical guide to creating a French garden in England, is addressed to "British men," who are visualized working on its pages.¹⁷ Clearly, however, the rise of historical market-gardening training programs for women indicates that they too sought to train in intensive culture before, during, and after the Great War.¹⁸

The record Cockerell and Nussey left behind of their garden, *A French Garden in England: A Year's Record of Successes and Failures in Intensive Culture* (1909), represents their gendering of an alternative agricultural experiment within a wider cultural and material exercise in growing local food in England. If, typically, garden books have a "strong whiff of the conservative about them," as certainly William Beach Thomas's introduction to McKay's text does, *A French Garden in England* sets out to link the authors' project to their "freedom" as "two women."[19] Cockerell and Nussey's book deploys the discourse of the garden diary to provide not only a guide to "modern gardening," but a record of a viable alternative life for middle-class women and the land.[20] A hybrid text of New Woman and early green print culture, it is addressed, like McKay's manual, to those who might experiment in intensive culture. At the same time, it participates in the early twentieth-century feminist discourse Matthew Beaumont and Rita Felski identify, providing, on the one hand, performative "evidence" of women's equality and the efficacy of early green principles, and, seeking, on the other, "to bring into existence through its own writing that political community to which it aspired."[21] Published by New Woman polemicist W. T. Stead, printed and bound at the Arts and Crafts press W. H. Smith & Sons in Garden City, promoted by Cockerell and Nussey's horticultural mentor, Lady Frances Warwick, and read with pleasure by reviewers as well as women like Jane Morris and Charlotte Mew, Cockerell and Nussey's book seeks to bring an equitable, ecological future into being in the present.

If in general, "the New Woman of the 1880s and 1890s was widely perceived, by both supporters and detractors, as an emissary of Nowhere," Cockerell and Nussey suggest utopia may be found on a plot of land two acres wide.[22] The women move determinedly in the text, rapidly transforming into their new profession as they travel swiftly by train or bicycle to their country garden. Buoyed by this newfound mobility, they follow a period trend in making their personal life "the crucial vector" of their emancipation.[23] Much is at stake in their venture, as their success or failure will undoubtedly become "proof" of the possibility of "alternative social relations" for women in modernity.[24]

From their introduction, then, the narrators infuse their record with future thought, playfully incorporating some of utopian fiction's "blend of romantic adventure story and social blueprint."[25] They introduce their village in Sussex, for example, from the perspective of Londoners traveling in a new country, reversing the trope of innocents abroad. "Here was rural life, indeed," they write, "a village with only three shops, each of which had its own peculiar fascination in that it contained so much of the unknown.... We had perhaps hit upon an undiscovered corner of civilization."[26] The details of their process of simplification—how not to tell the butcher they cannot "desecrate" the "sanctum" of their home with meat, for example—are explained as if these are necessary rituals, keys to the cultural practices of their new world. From the location of their garden, the topography of the future expands before them in a striking view: "Over the tree tops rising from the slope below, we could see miles of wooded country and the blue line of the South Downs swelling up to Chanctonbury Ring in the distance before us."[27] The place seems visionary, "enchanted," like a cottage in an aesthetic ballad where the "grass so green" seems to grow both inside and outside of the house.[28] This magical, anarchic setting is their immediate future: "We knew that here we should make our home."[29] All that remains for perfection is the replacement of a servant with "an imaginary maid of all work, called Betsinda," after the character in William Makepeace Thackeray's fairy tale *The Ring and the Rose*.[30] As they mediate the garden's imaginative and material dimensions, the authors speak in alternately earnest and ironic tones. Dismantling the conventional practices of the Edwardian household and farm or garden, they co-opt the heuristic language of the garden record for their gender experiment. Moving swiftly through their journey to "our home," they suggest that the future may be cultivated now. One needs only to follow their example.

As the women erode distances between culture and nature, fantasy and reality, the present and the future, the city and the country, they make way for something new: an ironic ecology. Such an ecology, Bronislaw Szersynski argues, may "value and proliferate 'impure' and vernacular mixings of nature and culture, new shared meanings and practices, new ways of dwelling with non-humans" and humans alike.[31] Its legacy might be "a living, evolving plurality of shared forms of life" rather than an epistemology that strictly separates culture and nature.[32] Cockerell and Nussey pursue such shared forms of life, juxtaposing their urban sophistication and imaginary maid with the willful soil, plants, and "creature comrades" of the woods and garden, redefining companionship between many kinds of bodies.[33] From the first page, Cockerell's illustrations visualize this ecology in the shape of a crystal-clear cloche, brimming with companion plantings of lettuce and cabbage, surrounded by two rabbits and two snails (see fig. 6). This assemblage sits above the large uncial which announces the "W" in the beginning of the book's first sentence, "We are two women."[34] The overall effect suggests "new shared meanings and practices" between women, between plants, between nations, between technology and rural England, all the while developing the women's "new ways of dwelling with non-humans" in the shape of plants, crustaceans, and animals. This gender experiment models the potential for human self-interest and ecology to mesh. Recent references to their garden in classic works on alternative agriculture,[35] as well as new awareness of the ecological potential of what is now called "biointensive culture" suggest that Cockerell and Nussey's garden, while of course not ecologically perfect, shapes a sustainable, food-productive paradigm now appreciated as antidotal to the high ecological and

INTRODUCTION

WE are two women, not very robust but accustomed to hard work, who started a French market-garden a year ago with a

Figure 6. "We are two women," from Olive Cockerell and Helen Nussey's *A French Garden in England* (1909). (Author's collection)

human costs of monoculture.[36] Moreover, their book claims the year in the garden as "our" time, women's time, certainly queer ecological time, in its resistance to the heteronormative discourse of the "ancient apple queen" or "handmaiden" "trapped in time" behind men.[37] Within the pages of their book Cockerell and Nussey, writing as two women, reclaim growing time from heteronormative time. As they share their year, they build resilience into their project and their partnership; the temporal dimensions of the book create a queer utopian sense of what "we" and "ours" means. These pronouns make up the structure of the book itself, which is organized visually and textually, chapter by chapter, around the women's plural pronoun in its first page and in titles like "Our Crops," "Our Marketing," and "Our *Ménage.*" This affirmation of the presence of the two women in the garden disrupts the idea that individualist and heteronormative structures are eternal. Simultaneously, the agency and human exceptionalism implicit in the repetition of the pronoun is held in ironic tension throughout the book with that of those other beings placed above the women's uncial on the first page: plants, animals, insects, and even soil. All together, this assemblage of human and more-than-human companions, along with their wills and appetites, creates a horizontal rather than vertical relationship.

Steeped in the culture of Arts and Crafts as well as the scientific and experimental discourse of the New Woman, Cockerell and Nussey's introduction of an imaginative and often tongue-in-cheek aestheticism and even occasional fairy-tale discourse evades the limits of the garden diary and realist narrative alike, engaging the imagination to make readers see the possibilities of a new kind of "enchanted" garden.[38] In this garden, soil itself is alive, while the Londoners live in the woods.[39] Simplification of housework is taken seriously and the women live without carpets and clutter. Cheese, salad, nuts, and fruit, meanwhile, all seem fresh alternatives to the time-worn and costly British diet, as Betsinda is a fresh response to a very tired Edwardian domestic ideology and the classed and gendered division of labor on the English farm. In contrast to the large fields and heavy labor of the latter, Cockerell and Nussey's small holding is their proving ground. Tweaking the sentimental ideal of the traditional English farm cottage, grange, or manor house, the garden business they name "The Bungalow" mediates between city and country, deploying the nearby railway to transport their produce. The women's focus on developing a list of customers to whom they ship baskets of seasonal produce anticipates strategies just now catching on among community-supported agricultural projects today. Cockerell and Nussey's garden suggests, however, that their ecology requires more than merely changing practices; it involves interrogating and challenging the gendered and classed assumptions and entrenched concepts that naturalize unsustainable patterns of land use.

On leased land with poor soil on an estate in Sussex, Cockerell and Nussey grow food on "only one acre."[40] The train, rather than any one local place, allows them to construct their own soil by providing a steady source of urban horse manure. Defying logic, the tiny garden holds "two-hundred lights" or glass-covered raised beds and "bell glasses as well," within which diverse plants were clustered intimately, rather than in the traditional prim rows of discrete plants.[41] With precious little capital to invest, the women are forced to work conservatively. However, keeping their garden and even their produce small-scale means they can grow a wide array of tempting vegetables quickly, as they minimize losses and soil depletion associated with larger-scale projects. "Not fastidious about soil," as they work in amended raised beds, the women hope their record will be useful to "others" seeking to venture out into spaces that need only provide sufficient sunlight and water to overturn generations of gendered agricultural practices.[42] The "working expenses" for Cockerell and Nussey's "French garden acre" are also small, "about £100," a budget they attribute to the fact that "we work ourselves."[43] As the women tell their story of their hard work in the garden, their tone is often light and comical, mixing work with play, almost magically transforming the domesticity of gardening and the plodding pace of agriculture into the urbanity and mobility of the New Woman.

In pursuing those, Cockerell and Nussey were clearly privileged, having many advantages, including education. Nussey at least, likely had access to the not insignificant amount of £500 per year that Virginia Woolf notoriously prescribed as necessary for women to pursue creative work.[44] While both had been working for some time, they seem as well to have received emotional and perhaps financial support from male friends and relatives, although this sometimes came at a price.[45] A paradox of their iconoclastic project, then, is that it, like William Morris's work, was likely helped by a good deal of economic security. While there were clearly brothers, fathers, and aunts to negotiate, such help enabled them to experiment in disentangling themselves from the trappings of middle-class domesticity and particularly from the classed and gendered status quo of the English farm, itself a contributor to the agricultural depression. Cockerell and Nussey therefore challenged the human culture of English farming, in which the "elite land owning class" was held back by "class divisions, poor labor relations and middle-class farmers' aversion to manual labor."[46] As Olive Hockin would say a few years later, the English farm was designed to benefit the "Maester" and the "Missus," and the human culture of monoculture itself had a great deal to do with the agricultural depression into which Cockerell and Nussey were able to intervene.[47] "Improved" single-crop English farms were both too big for farmers to work "with their own hands" and too unprofitable for them to reconceive in the crushing face of cheap imports.[48] While farmers modeled their work lives on "the genteel conservatism of the 'manor house'" rather than on models of "industriousness," their failure to explore innovation and cooperative work contributed to the agricultural depression *and* opened the door to women willing to work cooperatively and innovate well beyond the limits of monoculture itself.[49]

Cockerell and Nussey then worked against the grain of the latter, and, while it is difficult to know how many other such women existed, it is clear that work like theirs provoked anxiety in dominant culture before, during, and after the Great War. In 1897, for example, Grant Allen, in his infamous anti–New Woman novel *The Typewriter Girl*, depicts his heroine, Juliet, bicycling out to the countryside to seek work in an anarcho-communist market garden. The food grown there is represented as stunted, the labor miserable, and, while there, the heroine Juliet is sexually harassed by a dirty, Kropotkin-like figure. She departs quickly. As stated in the introduction, the 1917 *Punch* cartoon suggests intensive culture thwarts the stable order of things. The implication is that it challenges the sacralized, masculinist creed of the plough and the field. With her gamine glamour and her transformation of carpet and petticoat into soil, this alternative agriculturalist challenges the status quo on an elemental level. The fear that such women will make men obsolete bursts out after the war in texts like D. H. Lawrence's "The Fox," (1922), which features two stereotypically "odd," unmarried, and prematurely aging New Women alternative agriculturalists, Banford and March, who run a poultry farm in Berkshire.[50] The story takes place in the depleted and fallow England of the agricultural depression, when even the nation's rich orchards were in decline, as they had not been cared for or replanted since the nineteenth century. Food of all types is scarce, and the death of the previous farmer suggests the crisis of rural depopulation. Both Banford and March are artistic and enjoy their bicycle-riding more than farming: "Alone in the fields by the wood . . . they seemed to have to live too much off themselves."[51] Although a farmer, Banford declares she is "afraid" of "Nature altogether."[52] The women struggle to make a living from their chickens and are unable to catch a marauding fox, who is soon personified as a soldier, Henry. He "masters" the masculine March and kills the "weaker" Banford by hitting her with a dead tree, which he had felled.[53] Once the property of a prosperous farmer, the farm itself now seems dead as well; the countryside is hungry and the women fail at their endeavor to make a living feeding it.

In "The Fox," Henry hunts animals, "a great addition to the empty larder."[54] But his real prey soon turns out to be March: "He wanted to bring March down as his quarry, to make her his wife."[55] After Henry murders Banford, the cool March blooms. She is placed back into a properly gendered floral discourse and awarded "*the* flower" of heterosexual "happiness" (emphasis in original).[56] The next step is emigration, departing from the farm, which is stranded permanently in the gloom of the agricultural depression that had been ongoing for forty years. The only viable alternative for this is the abundant "West, Canada, America." There, March "would not be a man any more, an independent with a man's responsibility."[57] Cockerell and Nussey's text, in which they record their happiness, their successes and their failures, and the richness of their soil, speaks back to these clearly misogynist and homophobic texts, which can imagine neither two women happy together nor a green future for England. *A French Garden in England* is then antidotal to Lawrence, *Punch*, and Allen. Not only does it record in women's own voices their pains and pleasures in their experience of alternative agriculture, but it naturalizes the latter anew against the grain of Lawrence's terrifying biological determinism. Contrary to the stereotype of the tragic queer or "odd" woman without a future, as established by Allen and Lawrence, Cockerell and Nussey see their garden future as bright and their household as eminently satisfying. Indeed their text may anticipate other imaginative, queer, rural women's texts such as Sylvia Townsend Warner's *Lolly Willowes* (1926). Both may be understood as protomodernist reclaimings of rural space for alterity.

In making a record of their garden, then, Cockerell and Nussey record how they rise, transformed from amateurs into professionals. At the same time, they transform their land and even people through their garden, which shifts the desires and diet of their customers, creating a mutually gratifying garden-to-table network:

> We began to ascend step by step, and at the end of September, the end of a French garden year, we felt we could fairly claim to have reached the point to which we had hoped to climb, seven months only after we had sown our first seed. We had by then paid the working expense of the garden and had begun to get so many unsolicited demands from private customers that we began to feel quite like partners in an old-established firm . . . we steadily increased the number of hampers we sent out weekly. People liked their vegetables fresh—direct from a garden, and many of them wrote encouraging letters.[58]

In this passage, Cockerell and Nussey use the New Woman's narrative strategies of epiphany, plain speech, and the practice of recording a "female first" to argue that women may partner with the earth to the advantage of both. They construct their own alternative to emigration and marriage as well as a vision of a green *aube de siècle* that challenges popular anxieties about the New Woman and the "sense of doom" farmers experienced as they lost hope in monoculture during the great crisis of the agricultural depression.[59] In an age when large-scale agriculture is increasingly recognized as the largest single environmental polluter tied to concomitant environmental injustice, their work models not only gendered resistance to the latter, but ideas for sustainable alternatives now. Their work reveals the extent to which ecological problems are widely systemic, deeply rooted not just in ecological practices, but in cultural assumptions and privileges that themselves have taken the shape of Nature, like the image of Flora in a field of flowers or the "Maester" and the "Missis" amid waving fields of corn on the English farm.

Cockerell and Nussey's garden, however, like any utopian project that takes place on material terms, is "a vision rather than a dream," a complex record of material encounters, contingencies, and limits, as well as opportunities.[60] The former become more complex when the garden is fully historicized. While the women are seeking, like Mary De Morgan in her tales, to explore their own bodily autonomy, Cockerell and Nussey gain practical gardeners' sharp, sometimes costly insight into their ultimately horizontal, material relationship to the garden. Arriving there to prove their mettle as two women, they find they are not alone, but rather surrounded by a host of other actants also inhabiting their one-acre garden home: animals, insects, birds, weeds, chemicals, heavy metals like lead, and soil itself, which is living rather than inert. This coming to consciousness of their intimacy with materiality itself is characteristic of New Woman ecologies, an unexpected insight fostered by testing their own bodily autonomy. Rigorous self-analysis of such experience in the tradition of New Woman narratives ironically often results as much in seeing and understanding the ecological entanglement of all living things as in measuring women's performance, their success or failure. Results are sometimes unexpected. The ironic ecology Cockerell and Nussey construct takes the reader well beyond the limits of utopian narratives into some harsh realities. Although their garden grows to be "more successful than we could have dared hope," Cockerell and Nussey's story expands beyond the borders of their book to the conclusion of the garden itself, offering ecological insights entirely unexpected, timely, and disturbing.[61] As a study in both representation and potent material encounters across bodies, the complete ecological story of *A French Garden in England* explores both the pleasures and the pains of this intimacy.

Companion Planting and Creature Comrades

To be sure, *A French Garden in England* intervenes materially in monoculture and in the limits of the lady's garden book, as well as in the heteronormative, reproductive natures that such texts inscribe. The book is, like other New Women's texts, performative, writing women's capacity for labor, action, and experimentation into being even as the women record their failures along with their successes. The text includes every detail for success, covering what tools they use (they urge avoidance of "ladies' sets") and what they wore in the challenging weather of Sussex to remain mobile and keep dry and warm. Their illustrations of themselves at work in blouses, ties, skirts, tam o'shanters, and straw hats deploy a New Woman smartness that suggests their status as a new kind of middle-class professional woman. In the text, their descriptions of their work foster the idea that nearly any challenge can be met, given the right technology, clothing, or skill. Through their use of the "right" tools and clothing as well as through transparent demonstrations of how they gained skill through study and labor, the women reject the idea that women are essentially suited to gardening. Their "failures" are an important part of this agenda. Part of the interest and suspense the book creates lies in the women's admission of a lack of expertise and the consequent challenges that ensue. Text and image collude to make wry or ironic comments on their efforts. A dog-sized slug with a desperate woman chasing after it with a handful of salt, for example, playfully challenges the text's confident discussion of how best to eradicate slugs with salt "systematically." These challenges are followed by the text's triumphant final chapter, "Our Ménage," which ends with a full discussion of how the women, aided by innovations like a self-lighting stove, live and thrive in the garden surrounded by "Health, bright skies, congenial companionship" among "many creature comrades who lived with us in the wood. . . . Surely there is little else that heart of human can desire."[62]

From the beginning of the book to the end, this kind of openness sets the tone for text and image. The signal image of the bell glass epitomizes the women's candor, inviting the reader into the text with near-total transparency. In their garden both bell glasses and "lights" or moveable windows over the raised beds are crucial to their practice. Glass defies the seasons and the elements, creating adjustable microclimates that allow the gardeners to grow food for eight months of the year. Although the cloches were in reality "opaque as to prevent scorching," they are clear glass in the book's illustrations, enhancing, in the tradition of Victorian glass culture, the reader's access to the produce and to the narrative.[63] Visualizing produce often alongside the gaze of hungry creatures who seek the "tempting morsels" of fresh cabbage or cos lettuce through glass creates a playful barrier between the reader and the plants, inviting her to acknowledge her appetite and share the garden creatures' hunger and pleasure in looking not just at the plants but at the women's lives. This transparency is thematic in the text, and it whets the reader's appetite for the story of the garden itself, in which the women's work, often minutely narrated and illustrated, seems eminently accessible, and their story truthful.

A French Garden in England therefore provides a visual and narrative window into a new environmental story in which women, land, and living things form a vibrant assemblage that exists side by side with a vitally nonreproductive human partnership. This vitality and vibrancy is bodied forth in the text's celebration of the authors as planters, women in motion rather than things "in bloom" themselves.[64] The authors' work—transplanting, forcing, and companion-planting diverse groups of plants so as to benefit each—makes the garden's productivity dialectical, a constant negotiation between the categories of nature and culture. Not customary in England by the early twentieth century, the practice of companion planting, visualized in the text by the different types of lettuce in the cloche, defies the English standard of prim rows of one kind of plant. The jumble of plants pushed together in raised beds rather than spread out in a field seems a bountiful chaos, a productive green anarchy. The garden fosters plants on a continuum that values biodiversity, from the "volunteers" (wild plants like strawberries the women may harvest sustainably) to the recruits (plants subjected to "early French forcing"). As growers, the women attempt to orchestrate these relationships and interventions, underscoring in the narrative not only their professional prowess and the success of their same-sex team, but also their appreciation for the vibrancy of their garden alongside their human contribution. The birds, insects, and plants that arrive after their planting become "creature comrades," who contribute by eating insects and take some produce as their payment.[65]

While the women are extremely industrious, the point of their raised-bed style of gardening was to prevent the kind of back-breaking work of ploughing that was understood in intensive-culture communities to exhaust the body and the soil. Arts and Crafts thinkers, early green socialists, and anarcho-communists were committed to the idea that local communities should feed themselves. But they argued that the way people worked and what they ate should change. Mary De Morgan's article on collective work, for example, cites the important contributions of alternative agriculturalist Peter Kropotkin (1842–1921), whose book *The Conquest of Bread* argues against labor-intensive bread and meat and advocates a diet of vegetables. Women like Frances Warwick who mentored Cockerell and Nussey embraced the idea of intensive growing methods for vegetables, augmented by scientific knowledge and glass. Warwick sought to combine agriculture with an intellectual life and meant for the women's agricultural college she founded, Studley College, to have the atmosphere of "Girton or Newnham" and to link agricultural training for women to "economic policy."[66] A socialist influenced by William Morris, she espoused his idea that food production could and should replace factory work for both sexes and that it could of course provide work for "surplus" middle-class women. Kropotkin also argued that current farming techniques associated with monoculture were brutal and should be modernized: "Imagine a peasant bending over a plough, throwing badly sorted corn haphazard into the ground and waiting anxiously for what the good or bad season will bring forth . . . think of a family working from morn to night and reaping as a reward a rude bed, dry bread and coarse beverage."[67] The key to avoiding this brutal work was to eschew the plough, bread, and meat, and to garden on top of the soil and live as vegetarians, as Cockerell and Nussey do.

Cockerell and Nussey, working as partners, queer the gendered discourses of conventional agriculture by abandoning the plough, perhaps the most sacralized and masculinized Western agricultural tradition. In the Pauline parable, for example, men bring forth bread from the land as they bring forth children from women's bodies with their "seed."[68] An understanding of sex interlocks with a gendered understanding of the land through the plough, stabilizing the patriarchal land ethic in which women's bodies and the land are both static resources reserved for men. Such understandings "are an important point of conversation between queer and ecological politics because they reveal the powerful ways in which" heterosexual privilege is naturalized and alternative sexualities are edged out even as soil is depleted.[69] By the same logic, alternative practices are powerful tools with which Cockerell and Nussey queer their garden, naturalize the New Woman's freedom and autonomy, preserve soil, and construct alternative natures that support and naturalize the independence of women, soil conservation, and same-sex partnerships. In alternative agricultural discourses such as Cockerell and Nussey's intensive culture, particularly in their images and descriptions of their hands at work in living soil, the earth may be constructed as exuberant, vibrant, active, and resilient.[70] Cockerell and Nussey thwart the metaphor of the plough as they work their brimming acreage, and the work they do there results in a complex mix of emotions: doubt, pride, wisdom, insight, and humility. Cockerell's gentle, even loving, depiction of a woman's right hand lightly digging a groove in the soft composted soil of a raised bed with two fingers while her left hand deftly drops a seedling into the groove challenges the masculinist agricultural image of digging and planting as a backbreaking, manly effort made possible only through "mastering" the mechanical tool of the plough (see fig. 7). Gardening on top of the soft amended soil, in raised beds, Cockerell and Nussey

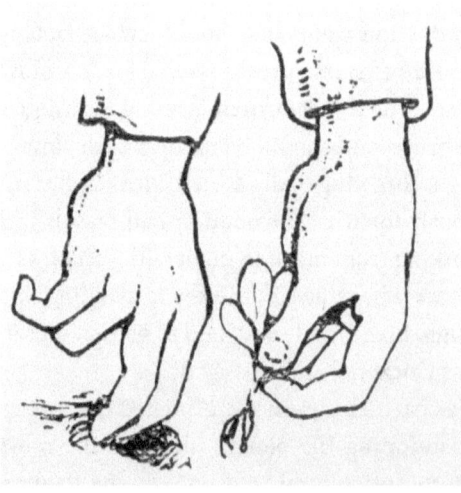

Figure 7. Planting seedlings, from Olive Cockerell and Helen Nussy's *A French Garden in England* (1909). (Author's collection)

avoid reenacting the violence of the plough. With their hands in the soil, they pose an alternative.

Touching the earth, water, glass, fertilizer, and even, more troublingly, the white lead they use to mend broken bell glasses, Cockerell and Nussey embody the "Ruskinian ideal of manual labor as a form of artistic expression and a manifestation of the worker's innate dignity" but challenge the gendering of that ideal as male and heterosexual.[71] If the gendered division of labor and the production of Victorian "culture was riveted to the materiality of their hands," then Cockerell and Nussey rework that materiality, defying the fetishized leisured hands of the earlier generation and extending women's hands well beyond the limited roles imagined for them by someone like Cockerell's godmother, Octavia Hill.[72] For Hill, ladies' hands are an authentic marker of Victorian women's elite agency. Hands hold power, they serve, they model superiority, and they rule. The housing settlements Hill purchased with Ruskin's support were to "bear the mark" of her hands.[73] Such is the primacy of the image of ladies' hands that Hill frequently mentions her hands and the experience of touch in her writings on her settlement work. So mentioned, her hands lend integrity to her work. But such is the prohibition against ladies' working hands that she uses them only metaphorically in the tradition of the household manager.

Cockerell and Nussey's representations of their hands planting, digging, weeding, mending jagged broken bell glasses with white lead (see fig. 8), and packing vegetables for market, then, are a strategic move toward gender equity and modernity, establishing their status as New Women who may, like the illustrator or typist, use their own hands to make a living. Such hands defy the rules. Thwarting her aunt's generation's resistance to women's utilitarian gardening and insistence upon women's supervisory and managerial rather than laboring roles, Cockerell's studies of women's hands, like hands in Victorian and Edwardian literature more generally, are important signifiers of the future, marks of evolutionary progress. In keeping with the paradoxes of hands in the same culture, they shift women into an authoritative new role in modernity as well as into a potentially anxiety-provoking, potentially exhilarating experience of intimacy and immanence. Moreover, as such hands do the work the Victorians and Edwardians outsourced to current or former colonies, they undermine the logic of empire and its unsustainable food miles, suggesting that the nation need not enslave or punish others in order to feed its own people. The women's hands in the soil foster an intimacy with their labor in their small garden that contrasts sharply with the atomization and alienation of labor associated with the expansive spaces of industrialized capitalism or agribusiness and the transformation of laborers into mere "hands" in those spaces.

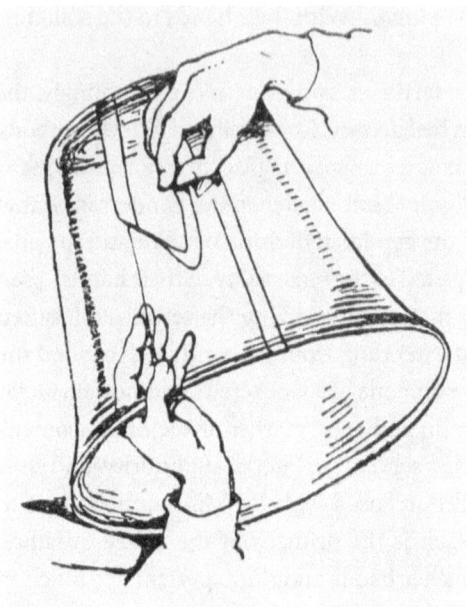

Figure 8. Mending bell glasses with white lead, from Olive Cockerell and Helen Nussey's *A French Garden in England* (1909). (Author's collection)

As a middle-class woman's leisured hands represented her family's transcendence of subsistence, her manual labor was a violation of Victorian ideals. In that imperialist "culture of mastery," the leisured woman was a sign of a man's wealth. Her presence marked their difference as a "higher" civilization, a chief sign of that ascendance.[74] The importation of food and England's "release" from the "backbreaking labor" of farming was also such a sign.[75] The nation's easing of the importation of food through technology like the steam ship was another, and its privilege to become a nation of gardeners growing solely flowers yet another. England's often female flower gardening then reified not just women's status as "blooms" themselves, reproductive beings about to be fertilized and then dependent upon men for subsistence, but was also an ecology that marked the culture's transcendence over need.

The authors of *A French Garden in England*, as growers themselves, thwart this gendered discourse and its assumptions about the very purposes of the gardening woman's body. Clearly, shocked reactions to middle-class women "spreading manure!" or "digging potatoes!" before and during the Great War respond to the sexual anarchy suggested by the middle-class woman grower.[76] The generational clash between Hill and Cockerell on the subject of market gardening centers around an unspoken fact regarding Cockerell's marriageability. Once ensconced in the garden growing food with Nussey, it becomes clear, surely, that any pretense at maintaining her niece's "bloom" and ensuring her accompanying biological destiny is over. Cockerell and Nussey's early twentieth-century publication, then, of an illustrated "record" of their garden that depicts their middle-class ladies' hands planting, digging, weeding, mending broken bell glasses, and packing vegetables for market is an important move toward their own declaration of bodily autonomy and sexual freedom within a new form of relationality—a female partnership on the land and in the home, both of which merge together. When glass cloches inevitably arrive broken from France, the mending is done inside so that the "the carpetless floor was thick with bell-glasses in every stage of break and mend, and we ourselves daubed over with a mixture of white lead, linseed oil and 'dryers' with which we had been" repairing them.[77] This invasion of the home with signs of work recalls the labor and life practices of the Arts and Crafts movement, where the studio flows into the household elementally.

Cockerell and Nussey work to make the strange, new, and foreign appealing in their text. This means accustoming their reader to the varieties and size of vegetables, like Grelot and Nantaise carrots, early lettuce, endive, cardoon, strawberries, peas and beans, cauliflower, radishes, and turnips, in shapes and sizes which were "not what we are usually accustomed to."[78] These smaller-sized, more intensely flavorful vegetables, and other produce like the starchy root celeriac, had almost entirely fallen out of cultivation in England, as had the practice of intensive culture itself.[79] But for their readers, one of the most controversial aspects of their garden is in the link between the produce's goodness and its cultivation in soil made from manure. As they state early on, their garden is unusual in England because it does not depend on the quality of the local soil. In fact, their garden is only made possible by a "quantity of manure, which used in the hotbeds when new, is dug into the land when old."[80] Anxiety about the presence of manure in their garden, however, allows the authors an opportunity to reframe soil itself. Their discussion is prompted by their anticipation of English anxieties about "the supposed injuriousness of vegetables grown with manure." Cockerell and Nussey refute the idea that growing vegetables in manure is impure and contest "the statement that 'Nothing on earth is half so pure as vegetables grown without manure,'" responding to "several customers [who] wrote asking us to explain our position."[81] To naturalize their work in their text, they counter the idea that food and waste are antithetical, or that the latter is a contaminant, and they recast waste as a crucial component of plants themselves, referring to it as "plant food."

The authors discuss their garden, then, at the cellular level, explaining in detail the relationship between soil and produce. Cockerell and Nussey take pains to show the roots of plants and seedlings, a feature that had long fallen out of the tradition of lady's botanical drawing, which focused intensely on knowledge of a plant's appearance and reproductive organs. Hands, roots, and soil, life-sized representations of plants with roots intact, and a thorough visualization and investigation of the importance of manure to food production are rediscovered for their chemical, tactile, material value. The authors offer complete disclosure of the science of their work, reassuring English readers of their competence as well as the goodness of their produce, as it builds their new subject position as female experts. Cockerell and Nussey's utopia, their ecological futurity, then, is built out of the smallest, most abject things, the goodness and importance of which they must communicate to others. At first glance, their produce might strike the reader as "as if the printer had made a mistake," but, they argue, its flavor and goodness is directly related to its small size and to the growing materials and process that they describe as nothing short of magical.[82]

To tell this story of conjunctions between unlikely things, "French forcings" in Sussex, manure and food, glass and plants, the women and the land, the city and the country, the authors strategically deploy a visual and literary discourse already associated with women's cultural authority, one in which Cockerell at least was well versed: the literary fairy tale.[83] The same discourse allows Cockerell and Nussey to explain the vitality of their garden soil, otherwise invisible to the wary English consumer. "The freeing of the nitrates from the manure and their preparation as plant food," they write, "reads almost as a fairy tale. Millions of bacteria, acting as scavengers, are ever busy with their beneficent cleansing work. All they require is lime with which to accomplish it. . . . It is when lime is deficient that the soil becomes foul, because the bacteria are thrown out of work."[84] This point is driven home through Cockerell's ingenious image of the bacteria as weeping laborers who throw their garden tools on the ground while a bucket labeled "Lime" rolls empty at their feet (see fig. 9). Turning the bacteria into despondent unemployed workers shaped as magical creatures transforms them into sympathetic beings, living things with their own needs. The emotions of the tiny figures make an environmental world of soil culture agentic, visible, and alive, and valorize the existence of what is normally unseen, derided, and trivialized. This transformative soil ecology explores how an inert substance becomes vibrant matter, excrement becomes "plant food," and plants become delicious human food brought into being in a partnership between two living things: human hands and dirt. That story is here an exhilarating tale, an intimate exploration of the most local of routes and patterns, from hand to soil to plant, from plant to table, from table to mouth and, inevitably, back again to soil. In explaining that the soil

Figure 9. "Bacteria Thrown Out of Work," from Olive Cockerell and Helen Nussey's *A French Garden in England* (1909). (Author's collection)

is dynamic and mutable, Cockerell and Nussey explore the idea that the environment is not a particular place, but is in the food people eat and in the ecological systems that make food and soil and people. Life in the garden is less a Darwinian struggle than a cooperative assemblage in which an exuberant variety of beings make up the stuff and substance of life.

As any reader familiar with Christina Rossetti's "Goblin Market" will know, foreign fruits and vegetables in the long nineteenth century were alluring but suspect: "Who knows upon what soil they fed/ Their hungry thirsty roots?"[85] Cockerell and Nussey reverse this fantastic and aesthetic visual vocabulary to enchant local food. A French vegetable grown in England is seen anew, dazzling under glass. Or it is viewed from the ground up, sometimes through the eyes of "Peter Rabbit," snails, or ducks that gaze longingly at or taste the narrators' produce. These images, in the tradition of the literary fairy tale, expand "the limits of the real world" and "affirm the truth of imagination."[86] To conflate science and magic is to uncover a vibrant world, making the otherwise invisible visible and in this case proving that what could be imagined could be carried out by "two women" in their garden.

Most of all, Cockerell and Nussey's playful deployment of fairy-tale discourse, like Mary De Morgan's, does not end in rescue by prince or marriage, but rather presents the possibilities of women's cooperative participation in a thriving ecology of labor and the promotion of both human and more-than-human vitality. At the end of the text, the women succeed in selling their produce and developing a group of loyal customers. They have plans for expansion into an orchard. As the text offers possibilities for middle-class women's hands and bodies to do more than merely the lightest outdoor work, so does it represent the material becoming of both their partnership and the garden as they reimagine the land as a future space, a nexus for the conjunction of agents seen and unseen all around them. As they record and visualize the process of making their garden in charts, maps, images, and lists, they move far too quickly to become the iconic and static Flora figures of earlier Arts and Crafts radical print culture. Because the ability of the women to live as "two women" is entangled with the particular way they grow food through intensive culture, their "record" suggests that the Pauline basis for heteronormativity is importantly cultural, and so mutable. The vital quality of their reflections is remarkable for its frank appreciation of a single-sex partnership rooted in a garden but treading lightly on top of the soil in raised beds, growing diverse produce, and cut "free" from both the "backbreaking" work of traditional agriculture and the equally painful limits of both monoculture and of heteronormative culture in which the two are always daughters, sisters, helpmeets, and potential wives rather than equal partners.

"Our *Ménage*"

By the end of *A French Garden in England*, it is clear that the women's experiment has been a success. Not only is their garden thriving, but their partnership seems to be as well. The book's final chapter, "Our *Ménage*," deploys a complex word that can signify everything from a marriage to a household economy or both. For Cockerell and Nussey, the term allows the authors to discuss their household and daily lives as well as their happiness in their garden home. Diet, clothing, and stoves become the material base of a joyful life. Their process of simplification has tied them together in an affective material and emotional partnership that the very project of the book invites others to consider. At the end of the text they share this conclusion, an insight into their renewed consciousness, an epiphany that ecstatically expresses their awareness of the rightness and completeness of their lives:

> Health, bright skies, congenial companionship, the visits of our friends, the sympathy and help of all around, the knowledge that we were making steady progress in what we had set ourselves to do; the many creature comrades who lived with us in the wood; the freshness of the dawn; the awakening of the birds; the smell of the mossy earth; the pushing of the growing things in spring; the moon behind the trees—the color of it all! And then the long winter evenings with a log fire and a pleasant book and the consciousness of a good day's work done. Surely there is little else that heart of human can desire.[87]

In this passage, the materiality of the garden comes together in a barrage of sensation to sanction the goodness of the women's partnership as an affective unit. The latter is witnessed by the color of the garden, the smell of the mossy earth, the dawn, the moon, and the reader. The affirmation in the last sentence is, like all aspects of their project, a small statement that speaks volumes to the preponderance of voices that would claim otherwise for a female partnership.

Instead, this collective is made a joyful presence, proven in text and image. The exuberant world around them of plants, birds, wind, sun, and their mutual commitment to each other as partners creates a sense of fulfillment and contentment that is striking in a rural garden that is often depicted—in pastoral, for example—as a setting for heteronormativity.

In contrast, the world that Cockerell and Nussey construct in their garden seems to support their partnership without question. Their ability to labor and live together as women "quite hidden in the woods," without the supervision of elders or employers, is a great relief. The completeness and fitness the women demonstrate is expressed in the text, but it is also bodied forth in the text's illustrations, which often show the women as competent workers. Cockerell's letters to her brother at this time indicate her happiness, admonish him to stop sending her money, and express the wish that he could meet her partner, Helen.[88] Moreover, in making the garden central to their lives, rather than ancillary to a primary domestic role, the women interrogate, as Mary De Morgan does, women's reduction to either symbols of "Flora" or handmaidens to men's work even within the progressive early green world of Arts and Crafts. While such an assumption about women seemed to have been allowed to pass unnoticed, they show that a different kind of garden forms a different gendered nature. In this sense, the narrators of *French Garden* invoke the kind of playful performativity Ann Heilmann analyzes in Sarah Grand's *The Heavenly Twins*, which "posits gender as an essentially fluid category, a performative act."[89] The narrators of *French Garden* represent themselves as characters in their own text, and they become agents of social change and proof of the existence and accomplishment of change itself. As New Women writers experiment with narrative forms like the novel, Cockerell and Nussey do the same with the narrative form of the garden manual.[90] They represent both the concrete action and the development of consciousness associated with urban New Woman narratives and express the completeness of their dual partnership within a powerfully unifying aesthetic and practical vocabulary and narrative. The achievement of the garden was a victory not just in gardening, but in reconceiving both home and labor, even as it had been defined in the bohemian and progressive worlds they knew.

However, if the women's bare hands on glass and in raised beds of soil are an invitation to a new experience of immanence, the latter also comes with risks. To read *with* the grain of Cockerell and Nussey's book is to see it as a literal, hands-on, queer defiance of monoculture and its companion gender ideology, patriarchy. But to read it somewhat against the grain, and from the outside in, is to consider the risks Cockerell and Nussey took in the garden, spreading industrial soot, for example, on their raised beds and using a mixture of white lead and boiled linseed oil to mend broken glass. The images of their bare hands on white lead drive home the knowledge of the toxins they were exposed to. Understanding these risks means understanding that touch works both ways. This reverses and challenges traditional ecological trajectories in which *human* bodies impact people, places, and things, leave their mark on them, as Hill says, almost to the exclusion of the reverse. In the discourse of the Anthropocene, for example, human hands "manufacture" the earth and sky.[91] However, the earth is also vibrant, while hands like the rest of the human body are also porous, able to absorb as well as to shape and present those elements to other parts of the body, nose, and mouth. The powerful discourse of human exceptionalism, even the powerful discourse of the New Woman striving for complete bodily autonomy, can obscure the extent to which bodies are matter and are impacted by matter. This slippage may be dystopian, but it can strengthen ecological thought and is demonstrated in the historical pains and pleasures of Cockerell and Nussey's garden, particularly in the garden's too early closure.

In 1910 Helen Nussey wrote an urgent letter to her landlord, cancelling her lease as "Miss Cockerell has cancer." Although the garden was in "excellent condition" the two women departed in haste.[92] The move was devastating, as the garden was at the height of its success and, certainly, it was Cockerell's "favorite spot in all the world." But the women were forced to take a flat in the city to be near the hospital. "We quite like our flat—for London," wrote Nussey, explaining to her landlord that she was staying in the city to "be with Olive."[93] Juxtaposing this news with the text's illustrations of the women handling soot or lead drives home Stacy Alaimo's theory of trans-corporeality, a recognition of the often unknown traffic between bodies, either human, metallic, bacterial, viral, or toxic.[94] One is forced to consider that human beings are far from separate from other actants on earth, and, once again, Helen Nussey's reflections on this issue are prescient.

In discussing her partner's illness, Nussey suggests a material risk at the Bungalow. She expresses concern that in "more than one case . . . it *seems* to have been the house which has given the disease."[95] Perhaps influenced by historical germ theories of cancer, or concerned about substances they had used in the house (it was there they used white lead to repair the broken bell glasses), Nussey insisted that she "disinfect" the house before any new tenants arrived: "With [the cancer's] origin so unknown it is impossible to be too careful."[96] Still, she attempted to get students to take on the garden from a nearby women's agricultural college. The school was on holiday and she was unable to make any arrangements. The idea that something was amiss in the lush garden home seemed difficult to process. By July 1910, Jane Morris lamented, Cockerell had died.[97] Neither Nussey nor contemporary readers could know definitively the cause of her death. Nussey's understanding that she "can't be too careful" about a substance or presence that she cannot see or identify characterizes the experience of understanding environmental risk, which remains elusive even today. Not only is too little known about environmental risk, but applied-knowledge systems, "disinfecting" the house, for example, are mere guesses at a solution. If "trans-corporeality often ruptures ordinary knowledge practices" causing "moments of confusion and contestation," Nussey's attempt to disinfect the bungalow is her attempt at "scientific mediation" with an "invisibly hazardous," unnamed entity. The result is a "disconcerting sense of being immersed within incalculable, interconnected material agencies that erode even our most sophisticated modes of understanding."[98] If there is any small consolation to the tragedy, however, it is Nussey's awareness of the intimacy between bodies of many kinds which contests the very notion of boundaries between inside and outside, nature and culture. Such questions challenge the concept of human exceptionalism itself.

The ecological story of Cockerell and Nussey's garden then underscores the idea that the environment is not a particular place protected elsewhere, but something that changes hands, shared between and within all things for good or ill. As Cockerell and Nussey's story evolves, it draws the reader's attention across established boundaries, from the impact of the women on the garden to the impact of the garden on the women. Their story erodes distances between vitalities that dominant culture would keep separate. As they mend glass in their garden bungalow, which seems "so green" that they joke the grass grows inside of it, their open hands challenge the division of nature and culture in disturbing and exhilarating new ways, questioning first a gendered hierarchy and division of labor, and then a vertical human relationship to matter itself. As "two women" they come close to the key insight of ecological thought: that there is a very fine line, perhaps no line, between the human and the more-than-human world.[99] In this sense, advocating for one is advocating for the other.[100]

In studying Nussey and Cockerell's intrachanges with the material world, even those that may have ended their garden, and particularly in considering the risks and consequences of their interaction with it, it is possible to site their environmental story as a "record" of more than they intended to write, a testimony of all bodies' entanglement with each other. This fact advances an ethic of partnership as the basis of environmental action for the good of all interests involved.[101] Cockerell and Nussey made their mark on the garden, but ultimately they suspected that it made its mark on them too, for good and ill. Their bold representation of bare hands on lead and glass and broken bell glasses on the carpetless floor disrupt the clean edges of ontological spaces and material substances such as flesh and mineral.[102] Haptic entanglement in the book is a radical vision of the reclamation of the gardeners' own immanence—her need, our need, to eat, and of the risks we take to do so. Also, they are a reminder that other unnamed hands elsewhere, far less valued than the authors' or ours, are feeding England, growing its food, transported at great human and ecological cost even now.

The final ritual through which Cockerell's family bade her farewell maintained the elemental intimacy of her work in her garden and in the Arts and Crafts movement. Her brothers scattered her ashes over the waters of Lake Coningsby, the same lake by which they had first gone to meet John Ruskin in their youth. There is no mention of Nussey attending the service. But it is affirming to know that her partnership with Cockerell at the "The Bungalow" marked Nussey as well, positively, throughout the rest of her very long life. Although she could not bring herself to return to the garden after Cockerell's death, she did spend the rest of her life advocating for urban environmental justice and food-productive urban gardens. She was to return to welfare work and to help found the London Garden Society. During the Blitz, she was instrumental in planning the evacuation of children to the countryside. She was also to write two more garden books and play a crucial role in keeping memories of her generation's environmental work alive. She enjoyed a wide circle of comrades, many of whom—like Eleanour Sinclair Rohde—were quietly responsible for reviving herbs in the twentieth century, thereby preserving English biodiversity before, during, and after the Great War. She would also continue to advocate for local food gardens. In the books and articles she would write, Martin Hoyles notes, "as early as 1939, Helen Nussey was sounding a warning" against the eradication of food-productive, working-people's gardens in London. She noted that the latter are "being swept away to make room for flats under our model housing schemes.'"[103] Even when those schemes were completed, Nussey argued for free, edible public landscapes, roof gardens, and bee hives on the tops of block-shaped apartment buildings, which to her seemed well-shaped for that purpose. This and more could be done easily, Nussey argued; all that was needed was "vision," a quality her generation had not lacked.[104] In the tradition of Arts and Crafts, Nussey argued that modern designers should look to the past for examples of garden cities.

It is now time to read the work of Cockerell, Nussey, and their cohort, and it is hoped that this reading of *A French Garden in England* is an invitation to do so. As Jane Morris noted, the book is a "delightful and wonderful production," especially "considering the circumstances."[105] It is also a rare gem, for not every market gardener, working sunup to sundown, can also find time and energy to leave a detailed written record of her work. For this reason alone, studying other graduates of women's agricultural and horticultural programs is challenging. Doing so, however, not only increases what we know about the New Woman on the ground as well as on the page; it also increases the far too limited archive of British nature writing. Finally, studying such women could be instrumental in understanding how ecological crisis may lead to innovation rather than simply despair.[106] "Ideas," wrote Nussey, "survive those who give them birth."[107] Reclaiming those ideas restores a gender gap in the understanding of twentieth-century ecological literature and culture, and it attunes us to how important it is to rethink monolithic narratives such as the Anthropocene itself, which may obscure small but compelling acts of resistance—struggles that may have gone unnoticed. For Cockerell and Nussey, cultivating one acre led to the rethinking of an entire gendered, economic system and the conceiving of a sustainable form of utilitarian gardening that might foster global environmental justice and food security.[108] My hope is that much more work will be done in finding more women "hidden in the woods" at the *aube de siècle*. These gardeners existed, even when they "pioneered modestly in unnoticed places" and perhaps did not necessarily want to be found.[109] After all, as Theodore Cockerell mused in a family letter, his sister Olive would be a great deal less worrisome were it not for "her *friends*."[110]

3

THE NEW LIFE AND THE NEW WOMAN

Utopian Ecologies in New Woman Writing

The example of Cockerell and Nussey prompts further study into the intersection between utopian ecologies and New Woman writing and culture. This nexus may be reclaimed in a wide range of much-neglected texts, expanding our knowledge of gendered ecologies in New Life literature and of utopian thought in New Woman writing. Memoirs, poetry, journalism, and fiction reveal that women who imagined or became involved in the New Life negotiated a gap between discursive understandings of the utopic New Woman figure and material arrangings of women as resources or objects. On the one hand, recovering women's representations and experiences of the New Life matters perhaps now more than ever, as it may do what utopia more generally does in alternative cultures: provide a "feeling of hope in the face of hopeless heteronormative maps of the present where futurity is the province of normative reproduction" and, I would argue, the linked project of industrialist capitalist environmental degradation.[1] On the other hand, understanding women's New Life ecologies may elucidate what has recently been called for: a truly "critical history of environmental discourse" that challenges "dominant extractivist and instrumentalist views of the natural world" and women.[2] The texts studied here reveal that the latter are sometimes hidden within alternative ecologies themselves.

   This chapter explores texts by writers such as George Egerton, Edith Ellis (Mrs. Havelock Ellis), Nellie Shaw, and Sarah Grand, all of whom wrote about real or imagined utopian communities that sought to enact versions of the New Life. The goal of the latter was to replace competitive individualism with cooperation and reclaim "the land for the people."[3] As Nellie Shaw recalls, historical anarcho-communist colonies such as Whiteway in the Cotswolds strove to practice gender equality and replaced marriage with free unions or fellowship, machines with hands, and mass consumption with "bread labor."[4] Individual ownership was overturned by the disavowal of money and property—a tenet Shaw and her comrades celebrated on the establishment of Whiteway by burning their land deed over a bonfire. Rational dress was de rigueur, and food, often vegetarian, was grown and shared, for "brotherhood is for every creature."⁵ For Edith Ellis, pursuing the New Life through what she describes as an "experiment[al]" cottage and farm venture confirms her belief "in work, in fellowship, in earth and in heaven" as well as in "the vigorous, sound Whitman man and woman of my ideal."⁶ Often referencing the latter poet as well as William Morris, Peter Kropotkin, Henry David Thoreau, Leo Tolstoy, Edward Carpenter, Havelock Ellis, and James Hinton (chief debater of Herbert Spencer's argument in favor of individual competition as the basis of evolutionary life), women also imagined and practiced the idea of bringing a green future into the present. Indeed, even as historical women in the New Life movement struggled with entrenched sexism and their heightened visibility in utopian projects, both a boon and a burden provided by the cultural work of the iconic image of the New Woman, they manipulated the latter, repossessing themselves through life writing, for example, to inscribe their future thought in print. Their writing of memoirs and historical articles complements their ecological acts of "survival, restoration, reformation, and reinvention" of unsustainable ecological practices.⁷ In the tradition of women's memoirs, their texts also "perform acts of repossessing the self and the world," even the utopian world long considered "a predominantly male domain."⁸ Finally, in managing discursive and material encounters in such texts, women also recall their own embodied acts of resistance. A struggle over water, laundry, and communism in Nellie Shaw's memoir *Whiteway: A Colony on the Cotswolds*, for example, reveals that in the New Life women challenged men to value all forms of labor and matter.

From the New Life, women report and imagine their own pleasures and privilege as well as their struggle with sexism and sexual harassment. Earlier Ruskinian "Simple Life" experiments, as Vicky and Fredrik Albritton explain, often excluded the full recognition of women's equality.⁹ As gender equality and the literature of the New Woman became widely accepted in New Life communities, however, middle-class women also asserted their claim to alternative futures. Reclaiming a year for her lover Lily Kirkpatrick in her *Lover's Calendar* (1912), Edith Ellis transforms the mundane into what Muñoz describes as "ecstatic time," seeing a "queer future on the horizon," one which embraces the New Life, sharply disassociates from industrialized capitalism and its gender conventions, and hence offers an alternative to "straight time."¹⁰ Indeed, referencing Oscar Wilde's "The Soul of Man in Socialism" in his epigraph to his work *Cruising Utopia*, Jose Muñoz may usefully invite ecocritics to think back to earlier temporal dreams. These may include women's *fin* and *aube de siècle* interventions into the straight time of the Anthropocene, in which "progress" means extraction and instrumentalization of the earth. Perhaps the ecstatic, imaginative, convivial, nonheteronormative, queer time offered by Edith Ellis's neopastoral *The Lover's Calendar* (1912) may be just the thing to reschedule our own? In current discourse, sustainable time imagines that straight time can be maintained, perhaps by a mythical "green capitalism."[11] This discourse too could benefit from a disruption from historical futures. The latter may intervene, in particular, in the purportedly "pragmatic," individualized and privatized, neoliberal agenda of contemporary sustainability that places women's consumption and recycling practices at a similarly "meager" apex of its future imaginings.[12] Certainly, retreating from women's utopian thought because of its flaws, Greenway notes, has historically done ideological work, reinscribing the impossibility of solving seemingly "intractable" problems such as institutionalized sexism and heteronormativity as well as "war, terror, oppression and environmental degradation."[13] The hopefulness expressed in Edith Ellis's favorite quote from Thoreau—"Love must be as much a light as it is a flame"—is especially valuable right now.[14] Studying her illuminating work is salutary, as it deepens an understanding of women's historical and cultural work against the grain of the Anthropocene.

Finally, as inspiring as such women's writing is, the work of reclamation is also useful in understanding the often hidden exclusivity of early social movements such as feminism and environmentalism. In particular, women's construction of the future may, through engagement with eugenicist thought, reveal a fine line between utopia and dystopia.[15] The ecologies to be studied in this chapter are alternately exhilarating and disturbing, encompassing a wide range of future visions, including queer, maternalist, eugenicist, and apocalyptic. The confluence of these last three qualities in Sarah Grand's *Adnam's Orchard* provides an early example of how ecological thought may produce a focus on particular bodies and places that matter. As the redux of Margaret Atwood's *The Handmaid's Tale* and the success of Sarah Hall's darkly imaginative millennial novel *Daughters of the North* disrupt our own myths of sustainability, these historical texts represent a tradition of women's resistance to the same. It is useful to draw a line back from the futures of Atwood and Hall to those of their literary foremothers, such as George Egerton's "The Regeneration of Two," *Whitewaym*, Edith Ellis's writing, and Grand's apocalyptic *Adnam's Orchard*. These women's historical and literary interventions, with all their ideological complexities, anticipate Gilead and Carhullan.

Historical Alternative Communities

Women's utopian imaginings in the nineteenth and twentieth centuries take place within a context of the establishment of a wide range of historical alternative communities. In *Alternative Communities in Nineteenth-Century England,* historian Dennis Hardy identifies ten major agrarian socialist or cooperative, anarchist communities formed between 1876 and 1899. Each of these was a working community, a "practical utopia" such as Norton Colony, the Purleigh Colony, Clousden Hill Communist and Cooperative Colony, and Whiteway and Wickford Colonies.[16] They followed the ideal of "back to the land" that espoused the rights of the British people to lands of which they had been "robbed ... in a shameful and wholesale way" during the enclosure of the commons in the eighteenth century.[17] As Nellie Shaw, a founder of the Purleigh and Whiteway Colonies, writes, the majority of members in such communities agreed that "the land belonged to the people, not the landlords, who by the jugglery of the law held it as private property."[18] Communities such as Whiteway followed Tolstoy's principles; hence "the first duty of man was 'bread labor,' to produce one of the three necessities of life—food, shelter or clothing. All of these were directly or indirectly the product of the land, consequently the land should be free for people to use and produce these necessary things."[19] In land experiments such as Edith Ellis's Cornish farm, the goal was to reduce consumption by simplification, living out "Goethe's demand that we should all live in the Whole, the Good, and the Beautiful."[20] In their "practical working-day life," some such communitarians sought to follow "Kant's dictum that every human being should be an end in himself or herself, and not a mere means to the end of another."[21] For the majority of land colonies, agriculture formed "the basis of all constructive work."[22] As Nellie Shaw and Edith Ellis recall, another primary objective of such colonies was the rethinking of space. The kitchens at Whiteway and Fellowship House, became a place for all to work and socialize. All doors of individual rooms were unlocked. Former rational dressmaker Nellie Shaw describes a communal men's clothes cupboard at Whiteway with clean clothing for all men to use. Initially, no property—tools, bicycles, clothing—was to be "private." Space was unenclosed, without gates or hedges. Meals were eaten outside, to reduce household labor and to intensify elemental contact with the earth. No one was to be head of the collective household and everyone worked.

In the 1890s, these decisions followed out of anarcho-communist rejections of the authority of the state and centralized control. During this decade, revolution was considered imminent, and preparation was considered both reasonable and transformative in and of itself.[23] Once revolution came, "the replacement of competition and authority by cooperation would have the purifying effect of a return to one's natural state."[24] Economically, a simpler localized production and distribution system would be established while "the wasteful services and distorted pattern of consumption under capitalism would be eliminated."[25] Thoreau's *Walden,* published in England in 1886, and Tolstoy's writing and simplification practices in Russia captured the imagination. Journals, clubs, churches, and associations fostered a revolutionary discourse that latched on, as Edith Ellis did, to these authors' words and example. These urban anarchists made much of Thoreau's "ideal of the fusion between body, soul and surrounding."[26] For them, breathing London's smog literally embodied their relationship to the state. In place of the factory, after the revolution people would return to the land. Edward Carpenter promised that the land "poised as in a dream" was "waiting for the kiss and the re-awakening" of revolution. It was "waiting for its own people to come and take possession of it."[27] Then "the broad waters, the air" would become "transparent."[28] This radical stewardship model sought to redress the loss of the commons and imagined the relationality of land and people as a mutually beneficial ecology.

Nellie Shaw, who belonged to the Fabian Society, the Croyden Brotherhood, the Independent Labor Party, and several other socialist groups, describes the 1890s as a period of intense reading, meeting, and talking: "The Fabian Society was in its palmiest days, following out its policy of permeation with much vigor."[29] The topics of the day included "Land Nationalization, Free Currency and other Socialist questions."[30] Having heard first-hand accounts of Tolstoy, reading Emerson and Whitman, and learning of utopian communities formed by groups such as the Shakers in the United States, Shaw and friends from the Croyden Brotherhood Church began to conceive a plan to go on the land after much communal discussion and "heart-searching" at the prospect of putting "the teaching of Tolstoy into practice."[31] Their first experiment, the Purleigh Colony, failed, suffering from class tensions and difficulties over celibacy, which Tolstoy encouraged. Doing away with compulsory celibacy, at least, Shaw and her friends quickly moved on to form a colony on forty-two acres long-named for the light color of its limestone bedrock, Whiteway, in the Cotswolds. There the community was to "thresh out" the difficulties of living as if the coming revolution had already taken place.

Like Edward Carpenter's home Millthorpe, in Sheffield and the nearby Norton Colony, as well as C. R. Ashbee's handicraft community in Chipping Camden in the Cotswolds, such places, although "little Utopia[s]," pursued personal transformation through simplification and work as an antecedent to large-scale revolution.³² In the case of Ashbee's guild, the social experiment resulted in objects of lasting value. In the case of Edward Carpenter, the same resulted in texts like *Towards Democracy,* a Whitmanesque prose poem that "speaks of the connection of bodies, things, places, actions, and feelings in networks that" became a future space demarcated as one in which "things *already* were otherwise."³³ For Carpenter and his friends, like Edith Ellis, this future space was demarcated in books, journals, handicrafts, farms, gardens, and in the restoration of old buildings and the construction of simple new ones like the communal Meeting Hall at Whiteway Colony. All of these indicate the protoecological nature of their work. But it is perhaps another materiality that makes their work demonstrably ecological: conviviality. The Whiteway communitarians loved *Walden,* but they never thought to be solitary, nor did they have any wilderness to which they could retreat. Their collective work is particularly timely for us, as it is only quite recently—after a century of environmentalism that shaped itself around highly individualized concepts of wilderness, heritage, conservation, and self-sufficiency, leading to equally individualized and privatized ecological solutions—that environmentalists have recognized that they "desperately need to learn how to celebrate community, too."³⁴ "A convivial environmentalism," Bill McKibben writes, "one that asks us to figure out what we really want out of life, offers profound possibilities."³⁵ New Woman literature and culture that consistently examines the issue of women's autonomy within relationality is well poised to consider how utopias bring people and the land into intimate but collective relation, without hierarchy, sexism, homophobia, or visionary individual leaders.³⁶ Other women, such as Sarah Grand and George Egerton, invoke the spirit of cooperation only to struggle mightily with its perceived flattening of the ideal New Woman they imagine in all of her intellectual, spiritual, and physical superiority. This chapter will study a dialectic between women's historical communities and their fictional counterparts as authors represent the intersection between the New Woman figure and the New Life.

"The Womanhood of Tomorrow": Edith Ellis and the New Life

Edith Ellis's writings on utopian communities express her long-term interest in creating cooperative, nonheteronormative futures through experiments in living. Ellis, who identified as an "invert," wrote under her married name, Mrs. Havelock Ellis.³⁷ She was a founder of the Fellowship of the New Life and the secretary of their Fellowship House, an urban experiment in simplification and communal living. In her article "Woman and the New Life," Ellis urged women to join the ranks of experimental communities. "The only way to bring the ideal into the real," she wrote, "*is to do it*. . . . *Now* and *here* as much as possible, in spite of failure, loss, and weakness."[38] Women "should live for the future as well as the present."[39] Putting aside "prudishness" and "dolldom," women should "steer into the deep strong waters of the Womanhood of Tomorrow."[40] In this future, men and women must be "equal comrades, shoulder to shoulder in the work and joy of life."[41] Ellis argues that "woman, *here* and *now* may be the chief herald or the chief hindrance of the New Life."[42] The participation of women in the latter, however, was, fraught with difficulty. Her novel, *Attainment* (1909), a roman à clef, is a comic rendering of her experience in pursuing the New Life at Fellowship House, as told through the eyes of the novel's heroine, Rachel Merton.

That novel is joined by a number of other writings that reflect on her time in Cornwall, where she had moved with her husband after leaving Fellowship House. In Cornwall, Edith and Havelock Ellis practiced a form of "semidetached" marriage that made room for both partners to maintain financial independence, live apart, and take other lovers. Early in her marriage, she took out a loan herself to acquire a long-term lease on a farm and cottages, beginning what she described as a "cottage experiment," meant to help support the farm. Ellis restored and leased the cottages to sympathetic friends and visitors, many of whom wrote of their stays in diaries and even in the periodical press. This venture was meant to test the theories Ellis later says she had only begun to explore in Fellowship House. Ellis's perspective on utopian thought can then be gleaned by exploring her roman à clef, her writing on her farm, and the intersections of the latter with her queer adaptation of the pastoral almanac form, *The Lover's Calendar* (1912), an anthology of love poetry that includes one poem for each day of the year. The book reshapes time as it gives the year over to love and the memory of Ellis's romantic affair with Lily Kirkpatrick, a painter from the nearby artist's colony of St. Ives, Cornwall. Cornwall also features heavily as an idealized landscape of the future in Ellis's novel *Attainment*, linked to the athleticism, fitness, and oftenremarked-upon good health of the heroine, Rachel Merton. Restless at the novel's opening, Merton initially rejects marriage as "an inevitable thing . . . like death."[43] She states that she is "too active to be content simply to dream. I want to live."[44] Her progressive physician father, remarkably, urges her to "go to London for a year or two and live as men live . . . unhampered and independent."[45] Such is her "natural courage and purity" that he does not worry about her virtue, but instead "wants [her] to fight a fight."[46] Rachel's clean-limbed physique aligns with the novel's invocation of the image of the New Woman and of purity feminism. This force, Ellis suggests, may play a role in revolution.

With, of course, significant differences (Ellis, for example, had lost both her parents and entered into the New Life without an idyllic country family to support and encourage her), the novel follows the course of Ellis's own life. Initially, Rachel, who has moved to London with her domestic servant, Ann, is caught up in a philanthropic scheme to help poor children vacation in the countryside. But it concerns her that in the household of the charismatic minister who runs the project, domestic servants cannot attend meetings and must sleep in the kitchen. She soon meets a William Morris-like figure, Robert Dane, a great socialist poet. As she sells him matches at a charity bazaar, he sees her incendiary potential, urging her to read Marx, which she does, in the British Library. As Rachel ventures away from the private philanthropic scheme and into the public spaces of London, she continues to talk with Dane, who explains to her the challenges of revolution. It is a difficult job, especially for women. "Women!" Rachel interjects, "What can women do? It seems to me we can do absolutely nothing in this stage of social evolution." Dane replies at once, "Everything! . . . You've got it all in you . . . the new life for us all."[47] Women can "hammer away at once . . . knock the stuffiness and unrealities from your homes tomorrow . . . practice statesmanship, and even new moralities there. You can hold back the most mystic force in the world if you choose, till we men are fit to receive it and until we've given you your just rights."[48] As Dane throws a match on the floor and empties his pipe on his plate, the revolution seems already to have begun. He clearly relegates women to the domestic realm, where they might shape the revolution through the nursery or, alternatively, in the absence of the nursery, by denying sex and children to their husbands. His words are "a tonic" to Rachel; they bring the future into the present. The role he assigns her in utopia is problematic, but the New Life is *now* and Rachel is full of hope.

Rachel's reaction to Dane jolts her from the dreary philanthropic venture in which she simply gave needy children a brief vacation from poverty. Now she is moved up and forward to revolutionary London. Time accelerates as that earlier "stage of evolution" she complained about is put behind her. As Rachel travels this new ground—going, for example, to a lecture with her friend Dane on "Progress and Solution"—the uninitiated note that they make a "strange couple."[49] A casual observer from outside their circle remarks: "It is life, and life always means love. They are in love."[50] But the observer is wrong. From this point until the very end of the novel, life does not "mean" heterosexual love but fellowship for Rachel. A lecture on "Progress and Solution" names her transformation. She "w[akes] up" and takes public space as a speaker, not a listener. "I've never spoken in public before," she says. "I feel that we are all brothers and sisters here. Even comrade seems too harsh a name."[51] Soon, with her new friends of both sexes, including a poet, a journalist, an anarchist, a distinguished botanist, and her domestic servant and companion, Ann, Rachel forms the utopian Brotherhood of the Perfect Life.

Rachel's excellent health and figure align her with eugenicist feminism as well as the iconic figure of the active New Woman of the future, who enters formerly male spaces such as utopia. Rachel's presence is elemental, associated with an almost apocalyptic fire, tasked with creating a new world and annihilating the old. A philosopher friend tells her: "I see you as flame. Flame cleanses, remember, and flings aside the rubbish it gathers round it when first it starts as fire."[52] As the fire in the grate burns, Rachel and her friends dream of the future, their minds drifting beyond the current time. If watching a fire produces reverie that leads to "invention and imagination," storytelling, Rachel now plots her future.[53] As the fire goes out, another is lit within her; she is "sick of theories and platitudes.... The great thing... is to do"; hence, Rachel forms the Fellowship of Perfect Brotherhood.[54] The collective will pursue "the whole, the good, and the beautiful.... Mutual aid... an ideal life, even under the present competitive conditions.... In Kant's formula, no member of the human family shall get his pleasure or development at the expense of another member.... Simplification of life... is the very essence of our scheme."[55] Women ask, and are assured, that "brotherhood" includes them. Both sexes will cook, clean, and cohabit without any one patriarchal authority over them. As they attain perfection, they will share their influence with the world. As a motto, they take a quotation from Whitman's "Song of the Open Road": "I will scatter myself among men and women as I go."[56] The words are printed on their letterhead to inscribe their collective future. What ensues is a lively portrait of the household, its comical challenges in the attainment of perfection at the most material level: eating, cleaning, sharing space, and coping with the intrusions of outsiders. The house survives its elite members' poor housekeeping, constant visits from "cranks" who seek to join and exploit them, the comical and ironic management of a vast array of dietary restrictions, and the men's mortification over scrubbing the front steps. Until the novel's end, it maintains the possibilities of Brotherhood House, eschewing and almost escaping the heteronormative marriage plot of the realist novel.

But what ends the future of the Brotherhood is the invasion of a very old and timeworn story in the shape of heteronormative sex. First, an enraged policeman shows up to accuse the elder botanist Brother Oram, "our very pillar of respectability," of harassing a servant next door.[57] The bobby hands a devastating piece of evidence to the incredulous group: a love note Oram wrote on the Brotherhood's engraved notepaper. The salutation of Oram's letter ("Dear Chickabee"), and its content (a pressing invitation to the young woman go to the music hall with the elder botanist), makes a mockery of Whitman's ideal, what was the "keynote to our humanity and general democratic feeling," and likewise, of the viability of nonreproductive, nonheteronormative fellowship of men and women.[58] The group comes sharply back from the future to the dreary present, if not the past. Indeed, the latter also catches up with another communitarian, Brother Busby, in the shape of a jilted, pregnant model whose portrait he painted. She reveals that when the painting was "finished and you'd sold it, you seemed to have done with us both. You'd got all you could out of us."[59] As Rachel leaves for Cornwall soon after, recovery seems unlikely. Brother Oram continues to act in an unbrotherly fashion to Ann; she reports in a letter that "Mr. Oram chucked me under the chin the other day and said, 'It's our turn next, Ann.' I should have flinked my duster in his saucy old face."[60] Soon Ann will be married, as will Rachel herself. As men's past and present indiscretions mesh in the novel, they rush the plot back in time to straight time, that "stage of evolution" that shapes men as predatory and biology as destiny for women. Rachel has now fallen in love and states that the earth is "an open," presumably heteronormative, "book."[61] Her biological children shall be "Nature's children."[62] "It is through them I shall know what attainment really means."[63] Rachel claims her agency from the same ground that objectifies her. As the novel ends, the sun is not rising, but setting.

How much utopian thought, then, did Ellis retain after Fellowship House? While Ellis's novel thoroughly critiques utopian experiments, it does seem clear that what she critiques most resoundingly about the Brotherhood is sexual harassment and the objectification of women, which obliterates the future in the present. It is clear that Ellis is "unable to describe a new kind of relationship between men and women as anything other than a fairytale," while "love between women is not even hinted at as a possibility."[64] However, Ellis's withholding of positive representations of heterosexual or queer utopias may simply put utopia back on the horizon rather than out of the picture. Moreover, her critique of sexual harassment in historical utopias is a relevant critique of why some historical social experiments broke down. Ellis's writing and life, however, make it clear that she had not given up on social experiments in living and that perhaps her next experiment, on her farm in Cornwall, became, for eleven years, the utopian experiment she had earlier sought.

Because either her husband Havelock Ellis or she herself destroyed her letters, we know very little about Ellis's farm and cottage experiment in Cornwall. While Havelock Ellis mentions the farm frequently in his autobiography, sharing details of its daily life, he is often dismissive and perplexed about her project. In one surviving article on the subject, however, Ellis again uses humor initially to distance herself from the earnestness of utopia as Greenway argues she does in *Attainment*.[65] But, even so, in "A Cornish Cottage Experiment," (1906), Ellis makes her belief in social experimentation and in her ideals of the New Life very clear. Her revival of the pastoral almanac genre through the anthology *The Lover's Calendar* also creates a New Life narrative, reordering the year as a "love history" between women that heightens awareness and care of the human spirit and the earth.[66] As a memorial to Lily Kirkpatrick, whom Ellis loved during her time on the farm, this text reflects upon the intersection of her exploration of ethical simplification and her experience of what she, after Havelock Ellis, termed her "inversion," or love for women.

Although his wife's farm seemed to Havelock Ellis an impulsive venture, Edith Ellis's gravitation from the urban Fellowship House to working the land parallels the move of other communitarians such as Nellie Shaw from the city to the country in pursuit of the New Life. In her article, Ellis clearly shapes her identity around the farm and cottage project, which involved the concept of "bread labor" (earning one's own meals), equitable relationships among all participants (animal and human), and the restoration of formerly neglected rural cottages. Ellis's article is written in part in response to an earlier article written by a guest whom Ellis perceived had romanticized her venture, making simplification seem both immensely profitable and easy. In a seriocomic tone, then, the subtitles of the article state the opposite: "DRAWBACKS AND REWARDS IN A PIONEER REALIZATION OF EQUALITY. FRATERNITY. AND LIBERTY—CIGARETTES WITH THE MAID—WHY THE HOUSE-LETTING EXPERIMENT FAILED TO PAY EVEN INTEREST OF THE CAPITAL SUNK."[67] These subtitles lead the reader to expect that the cottage utopia will come crashing down as the fictional Brotherhood of the Perfect Life does. However, by the end of the article, quite the opposite is true. Ellis affirms that the farm became a successful experiment in putting the "Whole, the Good, and the Beautiful" into the daily details of farm life through "the very commonplace way of not watering milk, not increasing the weight of butter by coarse salt, giving a fair lodging for a fair price, paying good wages, and treating and feeding men, women, and animals as men, women, and animals should be treated and fed."⁶⁸ While Ellis begins her article with comic complaints of material loss, she ends it with a story of spiritual gain: "There have been words spoken and deeds done during these years of labor and trial that to me are priceless. They have made me realize, as no mere theory could, that for employer and employed alike to live in any degree in the Whole, the Good, and the Beautiful, even for an hour, involves self-mastery first and self-giving next. Capital which gives a spiritual interest of this kind is an investment of energy and idealism which stands one in good stead for the next piece of training which comes to hand."⁶⁹ By the end of the article, Ellis's seemingly jaded perspective has been transformed into an entirely earnest one.

Ultimately, Ellis's article restates her New Life values and revives the possibilities for utopia. She places her work on a timeline in which the urban Fellowship House was just a prelude to her Cornish experiment: "Twenty years ago I was working out ethical ideals as the secretary of a society in London which had as its motto William Morris's famous saying, that 'fellowship is heaven and the lack of fellowship is hell.' This exhortation," Ellis continues, "was seconded by Goethe's demand that we should all live in the Whole, the Good, and the Beautiful, and the practical working-day demand of our society was Kant's dictum that every human being should be an end in himself or herself, and not a mere means to the end of another."⁷⁰ Ellis explains, however, that the farm took her further than Fellowship House. It was an "attempt to put into practice what, in those days, I believed in theory."⁷¹ After eleven years on the farm, Ellis explains that she "never really regretted the step I took when I left the ranks of the theorizers of the simple life in order to test its worth in a practical manner." What enabled the experience, she writes, was that "I believed in myself first and foremost, and I also believed in the vigorous, sound Whitman man and woman of my ideal. I believed in work, in fellowship, in earth and in heaven, and what more capital could be necessary for starting an experiment where equality and fraternity were foregone conclusions?"⁷² Ellis's article expresses deep satisfaction with her experiment. On the one hand, her ethos is markedly authoritarian; it is clear that she was the "employer" supervising "employees," and that her definition of socialism is perhaps reductive: it is defined in part as a former professor's daughter washing clothing and her own pleasure in sharing a cigarette with "the maid." However, what is absent from the narrative, from her perspective, is sexual harassment and sexism. In fact, Ellis lauds men on the farm who will "turn nurse" for a sick calf. She envisions the possibilities for men and women to overcome a present that sustains inequity and arbitrary gender roles. The collective she discusses eradicates cruelty to animals, the decay of local homes, and distinctions of class and gender in the present. Part of the purpose of her writing back to her visitor is to correct her incipient notion of perhaps an earlier version of "green capitalism." Ellis makes the point that her experiment seeks a more valuable type of capital.

The convergence of Cornish ecstatic time and utopia in both Ellis's "A Cornish Experiment in Cottages" and *The Lover's Calendar* (1912) suggests the possibilities of reshaping the heteronormative, ecocidal present into mutually loving relationality between people and more-than-human-nature. The calendar charts the birth of Ellis's love for Lily Kirkpatrick and the latter's death, listing one poem for every day of the year, every season of "love's history." This traditional pastoral genre, full of nontraditional poems, naturalizes her love for Kirkpatrick, who died in 1905. *The Lover's Calendar* imagines months of burgeoning desire, green passion, mellow pleasure, and tragic death. Poems that express these states of being replace the traditional calendar's purpose, its authoritative division of the year into days of the week, things to do, or national or religious holidays. Assembling a host of New Life writers such as Whitman, Emerson, Carpenter, Thoreau, and Morris, Ellis equates falling in love with falling in with their New Life principles. Thoreau's stipulation that "Love must be as much a light as a flame" appears at the earliest stage of Ellis's "love history," suggesting that the lover must not love blindly but must see her beloved's moral life.[73] Her interests must not be materialist, but noble. Next Emerson advises, famously, "Give all to love," and Carpenter, quoted from *Towards Democracy,* urges the lovers to make their love transform society: "Seek this act and that act and thousands of acts whose end is love—So shalt thou at last create that which thou now desires."[74] Relationality and intimacy replace dualisms, as suggested in William Penn's words: "Love is indeed Heaven upon Earth."[75] The selected poems continuously erode the distance between heaven/earth, body/mind, eternity/daily time. In March, Wilfred Blunt writes "thy much were Heaven: thy little Earth shall be. If not Eternity, then Time be mine, the human part."[76] Ellis creates a queer ecology in the poem as she naturalizes her love, comparing her to flowers, herself to "the cool wind that's blowing from the sea, . . . in the dew on the grass is your name, I the leaf on the tree."[77] By December, she places Tennyson's "In Memorium" near the end of the calendar: "I seem in star and flower to feel thee some diffusive power, I do not therefore love thee less. . . . Though mixed with God and Nature thou, I seem to love thee more and more."[78] Ellis's queer ecology celebrates the relationality of all things, the entanglement of life and death, the cycle of life and systems. All of this, moreover, is done in the utopian tradition of fellowship, among the New Life community represented in the anthology. Finally, as Jo-Ann Wallace notes, the calendar also includes generous white space on each page, in the tradition of the almanac or calendar, encouraging readers to write their love into the year as well.[79] Hence, the calendar becomes a convivial, cooperative project. This gesture seems particularly utopian and anarchic, inviting the reader to join Ellis's project as she trumps state, religious, or economic authority. The progression of the seasons marked through pastoral poetry, in fact, disassociates the reader from the passage of time as measured by days and weeks and returns it to the seasonal cycles of the earth and the birth and death of living things in the more-than-human world.

As a memorial to Lily Kirkpatrick, *The Lover's Calendar* is also a memorial to a historical moment and community, the New Life ideals and the time Ellis and Kirkpatrick spent together in Cornwall while Ellis was developing her farm and cottage experiment. Surely constructing a queer utopia, the poems in Ellis's *Lover's Calendar* intersect with her writing on her farm in Cornwall to suggest that loving Lily Kirkpatrick in this particular coastal place was both natural and key to her spiritual development. "A WOMAN—whose soul renders the common air sweet," writes Oscar Wilde for February 23rd in the calendar, "makes what is spiritual seem as simple and natural as sunlight on the sea."[80] Yet, while the poems in the calendar are often bucolic and full of nature imagery—particularly, as Wallace points out, of lilies[81]—they do not so much describe literal time or events shared with Lily Kirkpatrick on the farm as they figure the time on the farm with Lily in Christina Rossetti's words as "the birthday of my life."[82] While Ellis feared naming her love overtly, the meeting of her "Cornish Cottage Experiment" with *The Lover's Calendar* suggests that she continued to contribute to the cultural production of nonheteronormative, decentralized utopia in practice as well as writing. The intersection of both texts, moreover, encourages her readers to follow her anarchic path, literally and figuratively encountering the future of love alongside the more-than-human world, writing their own ecstatic time on the white pages of *The Lover's Calendar* themselves.

Clothing Utopia: Nellie Shaw, Whiteway, and a Colony on the Cotswolds

The reader of Nellie Shaw's *Whiteway* encounters two images early on in the memoir. The first page of the text reproduces a handwritten letter from H. G. Wells, who describes the "documentary value" of the book as "indisputable."[83] This is followed by a foreword written by a former communitarian, Joseph Burtt. Finally, as the first chapter begins, the reader finds another image, a recent photograph of Shaw herself. Captioned "THE AUTHOR," this image shows Shaw in her maturity, looking confident and relaxed. She sits in what looks to be a handmade wooden chair designed with Arts and Crafts simplicity. The austerity of the chair and the plain wooden siding of the structure behind her, however, contrast strongly with her intriguing attire. Shaw wears an elaborately printed or embroidered head scarf from which descend two heart-shaped pendants. Around her neck, she wears a large-scale, clearly handmade necklace of open-work silver or metal that ends in a pendant neatly framed by the neckline of her blouse. Shaw's expression and creative clothing produce a portrait of the accomplished anarchist as a mature woman. Cutting through H. G. Wells's praise and Joseph Burtt's romantic memories of Whiteway as "sweet Arcady," Shaw asserts her ethos by curating her own image.[84] Considering this assemblage of materials at the beginning of the text reveals the gendered importance of clothing utopia both literally and figuratively in Shaw's memoir. As a designer and maker of rational dress, Shaw confirms Christine Bayle Kortsch's claim that "New Women" might "sew a new world" through the "revolutionary potential of dress culture."[85] That latter gendered imperative intersects in her memoir with the historical inclusion of clothing in the sacred trinity of things from the earth—food, clothing, and shelter—that Tolstoy, and hence Whiteway Colony, identified as basic human entitlements under communism.

In her introduction, Shaw explains that she needed the support of a respected male author such as Wells in order to get her book published. At the same time, she makes very clear that *Whiteway* is her book by including her portrait. Constructing and managing her own image, as she had in the 1890s when Whiteway was formed, Shaw leverages the authenticity of the latter to enter herself into the public sphere of print culture in the tradition of the New Woman. Throughout *Whiteway*, Shaw recalls people's unconventional approach to clothing and appearance in detail as an integral part of the community's radicalism. She describes men in sandals and knickerbockers with "bronzed, gleaming calves," women with "cropped hair," and men "hatless" with "rather long hair," as well as an anarchist friend's "bright color scheme" that included his "emerald green flowing cloak" worn into town to the "amazement" of the villagers.[86] These "modes and manners" and an "early gaiety and carelessness as to what people thought of our manner of dress and habits" mark high points of utopian time and place for Shaw.[87] Clothes play an integral role as well in the "break-up" of communism during the colony's third year.[88] Shaw's attention to the meaning of clothes and their materiality marks crucial moments in her memory of the utopian project. She recalls color, cut, fabric, trim, damage, dirt, and laundry in some detail, stitching the latter into the fabric of the memoir, which records the very materiality, fragility, and ultimate resilience of the community.

At the outset of *Whiteway*, Shaw recalls that, in Croyden, she and her likeminded friends believed that "some kind of co-operative system [would soon replace] the present capitalistic system and commercialism generally."[89] Their land colony, Whiteway, would be a bit of the near future in the present. Initially, Shaw and her friends had formed the Croyden Brotherhood, which developed the journal the *New Order*, the Brotherhood Store and House, Brotherhood Dressmaking, and a cooperative laundry begun in one of the poorer sections of Croyden. It was in the laundry, Shaw explains, that she and the Brotherhood members gained "much useful information . . . at first hand as to how the poor live."[90] The key members of the Brotherhood are described as John C. Kenworthy, Arthur St. John, G. D. Blogg, Frank and James Henderson, "Mary Glover, Theosophist, Fabian and strong Suffragist, and myself."[91] Many of these figures pooled their resources to purchase the land in the Cotswolds, where they would disavow private property and embrace collective living. In August 1898, "a little group of two women and six men" moved to Whiteway.[92] Once there, Shaw recalls, the group relished the pleasure of shocking the local villagers with their modern clothing, through which they might defy Victorian sobriety and its unsustainable ecologies. Growing their own food not to sell but to eat, they refused to engage in the money economy. The comrades embraced gender- and class-bending attire as they pedaled, foraged, and planted for the future. Shaw, in short knickers and accompanied by at least one male comrade flaunting a dazzling emerald-green cape, had no time for extinction narratives but was dressed to defy the Anthropocene. Modeling an ironic environmentalism, Shaw took surfaces like clothing and the earth very seriously. As they settled at Whiteway, the group rejected the concept of private property. The land was claimed as future space, redefined against the surrounding villages as nonreproductive, atheist, Tolstoyan Christian anarchist or theosophist, communist, radical, feminist, sexually diverse (including the celibate, homogenic, and heterosexual), foreign, mobile, mutualist, social, and open. Taking possession of a plateau with very few trees, Whiteway became highly visible to its neighbors, who sometimes could not avoid observing, and sometimes actively sought to observe, the colonists. The colony was entirely open, without gates or fences, and women in particular were targeted by sightseers or photographers. Clearly, as Shaw writes the memoir, she is writing back to the media that misrepresented her purpose and her image at Whiteway.

The communitarians at Whiteway worked very hard. With very little experience in agriculture, they struggled and eventually succeeded in developing flourishing gardens, a bakery, and a dairy. However, their visual images and fashions were often what sold newspapers. Shaw, then, takes pains in her memoir to explain that "picturesque" men and women were often "dress reformers and had well thought out reasons for their style of dress," making a "convention of unconventionality."[93] Shaw herself took her work in rational dress seriously, and when she resigned her job at Goodwill Dressmakers of the Croyden Brotherhood, she published a farewell letter in the *New Order*, explaining that "the work done has been of an educational nature, and we have made dresses with an idea both to utility and beauty, on rational and hygienic lines."[94] She also enjoyed wearing "unorthodox but undoubtedly practical clothing, she designed and made herself."[95] Including photographs and detail of herself and her comrades in such unconventional attire in the memoir captures clothing's power to flout heteronormativity and sexism. Fabric, color, and design cut the colonists apart from the village and stitched them firmly to a utopic future.

For example, while "land hunting" in the 1890s for the location of the colony, Shaw and her friend Arnold Eiloart enjoy the confusion they cause as comrades riding bicycles through small villages from Croydon to Gloucestershire. "We were certainly a very unconventional pair. I was wearing what was then termed rational dress consisting of knickers and a neat Norfolk jacket reaching to my knees . . . my companion gradually divested himself of various garments, till all he wore was a short-sleeved vest, red braces, knickers and sandals." Shaw continues, "It is hard to say which of us attracted the most attention, but I think I was that one. The most opprobrious remarks were hurled at me, for the 'land girl' not yet being evolved, women's legs were still considered to be matters of secrecy."[96] As a promoter of rational dress, Shaw deploys both the mobility and the iconic image of "the New Woman on the move" to communicate that her presence is a sign of future shock.[97] In retrospect, she relishes her own progressiveness, knowing that she "evolved" first and women in breeches appeared years later with the wartime land girl. Ironically, while the rationally dressed New Woman figure evokes a mainly consumerist discourse, in making their own clothes—and sometimes in taking them off—the Whiteway colonists engage the discourse of fashion even as they reject consumer capitalism. When the local people are scandalized by colony attire, it serves the communitarians' utopian purpose of announcing the arrival of a new order.

The Whiteway group's sartorial invasion of the Cotwsolds is innovative on several levels, thwarting ecological strategies of sustainability or localism that advocate for "a culture that is native to a place."[98] In contrast, the colony advocates for the construction of a future place in direct defiance of an existing one. Whiteway is a utopic ecology that is shaped by an idea rather than a location. Shaw is intent upon discussing the colony as an anomaly in the local landscape, a place off the map of the region itself, with shockingly little interest in the area's history or in its sufferings during the agricultural depression or later during the Great War. These were many. Jane Bingham notes that local diaries such as *A Cotswold Village* (1898) or *Rain and Ruin* (1900) make "grim reading," recording "daily struggles with poor soil, terrible weather and disastrous harvest" from 1875 to 1900.[99] As thousands of Cotswolders left the countryside for town life, the Whiteway colonizers entered as proud "foreigners." Paradoxically, they went back to the land without any pretense or desire to adopt local farming methods. All agricultural neophytes, their work took time to master and they were often hungry. One spring, for example, they had only dried peas to eat and these "it was impossible to render soft . . . having been harvested too late in the season."[100] Shaw does not report asking for help from local farmers or studying their methods. Instead, the colonists learn from experience, eventually producing enough food to live on. Even when hungry, they decline to "mix with the villagers," preferring to stay "on the top of the hill, two miles away, . . . grow[ing] the corn and staple foodstuffs for . . . ourselves, and enjoy[ing] life on a table land eight hundred feet above the sea."[101] Farming the land at Whiteway, even badly, is prized as the colony's most politically resonant work; Shaw notes her own contribution with pride. Even as she reports this, however, Shaw often demonstrates a paradoxical lack of awareness of how the colony domesticates her. While it seems clear that all recognize the political meaning of growing food, men do not afford cooking, sewing, or laundry the same high status. By the spring of 1900, for example, Shaw reports that "the land was got in order and many kinds of seed sown." She adds, "In this work we three women took part, glad enough to get away from the kitchen out into the open air."[102] Shaw's memory records a classic double burden of traditional and alternative work. This makes her representation of a looming crisis over communism in the colony's third year all the more interesting.

To honor their Tolstoyan tenets, Whiteway made food, shelter, and clothing freely available to the community. Hence, they kept a communal clothes cupboard, which Shaw describes as a service the women provide for the men and visitors. The upkeep of the cupboard is important work Shaw finds gratifying, as it calls upon her sewing skill, her interest in fabric, and her ingenuity with donated material. Mending and sewing together with the other women at Whiteway also seems to have provided them with needed community time. The cupboard full of clean, well-crafted clothing, often put together through donated clothes and fabric, seems for Shaw to house the ideals of the commune. Maintaining the cupboard, then, ought to be perceived as crucial political work. The free clothes articulated the colonists "splendid attempt to create a little Utopia in the midst of a capitalistic world."[103] Not only did the communal clothes cupboard represent the sharing of material things at Whiteway, but it also represented the utopian ideal of the radical rethinking of space there as well. The clothes cupboard was unlocked, and people could enter and take clothes as needed, without prohibitions, boundaries, monitoring, ownership, or rules. However, the maintenance of this aspect of utopia was distinctly feminized and it compounded the overall labor for the women. Moreover, when visitors absconded with the clothing, they violated the ethical basis of the cupboard, and "this annoyed the women very much."[104] In the third year of the colony, the washing of clothes remained their last communal activity until it ended in a final, gendered dispute.

Shaw explains:

> The three women were quite willing to wash the clothes for the whole community as heretofore, provided always that sufficient water was fetched and wood gathered for the copper fire, which they expected lighted on Monday morning ready for the washing day. Certain men had hitherto collected the wood for this purpose—Francis [Sedlak] for four months, and [William] Sinclair for nine months—at the end of such time he mildly suggested that another man might take on the job and give him a rest. However no one would volunteer, and after much argument we women were assured that whatever was necessary would be done, and there was no need for us to worry about it. The next Monday morning when Lucy and I arrived upon the scene with our bags of soiled linen slung over our shoulders, we found no fire, no water, no wood; only a big pile of dirty clothes lying about.... From that time onward each man had to wash his own shirt, and so disappeared the last vestige of communism.[105]

Shaw's vivid memory of the men eating breakfast in the kitchen as the women confront them over the unlit fire indicates the men's great difficulty in fully accepting "women's work" as political work on a par with land work.[106] It is intriguing that a chief influence over the Whiteway Colony, Henry David Thoreau, despite his fervent embrace of self-sufficiency, also quite notoriously opted not to wash his own clothes. Instead, he brought his laundry home to his mother, Cynthia Thoreau.[107] Thoreau's laundry has perhaps done more to discredit utopian thought than almost any other historical or philosophical fact of Thoreau's life. However, the gendered logic of Thoreau's depositing his laundry with his mother and his exclusion of this labor from his radical

Walden ecology is rarely analyzed. Shaw's discussion of how laundry caused the "Break-Up of Communism," is, in contrast, the central event in her third chapter of the same name. Her detailed discussion of the dispute concludes with an unflattering portrait of the men after the breakup "avail[ing] themselves of our washing water, eagerly watching until we had finished, then darting out and putting their things in the tub."[108] Shaw essentially washes the dirty linen of masculinist utopia quite publicly, critiquing men's instrumentalizing of women and their failure to honor clothing even as they claim radical ground through their intimacy with the earth.

Indeed, while Whiteway challenged vertical relationships between people and land, this episode shows the conceptual limits of a gendered ecology in which men preserved a hierarchical relationship between *things*. Although providing free, clean clothing to all was ranked in Tolstoyan communism as important as providing food, and, although it was done on the same ecological basis as both were things of the earth which should be free, maintaining and caring for clothes and carrying water to wash them was abjected while tilling the soil was valorized. In confronting the men as they are engaged in the important communal ritual of eating breakfast, Shaw points out that their communism falls short of its ideals. Taking laundry seriously is a political act that necessitates carrying water for others. When Shaw notes with particular disappointment the appearance of "no fire, no water, no wood; only a big pile of dirty clothes lying about," she cuts to the heart of the men's attempts to disclaim their own materiality. Their abjecting of the latter is driven home in the "big pile" of dirty clothes "lying about" on the ground (the washing was done outside). Shaw shrinks from including some material in *Whiteway*, believing that her first draft had gone too far in saying that which should not be said. We can only guess what those things are. However, she does not shirk from detailing her own "breakup" with communism in this scene that involves the majority of the men's unwillingness to understand that clothing utopia is a political, if not a sacred, act. In fact, the episode becomes a crucial moment in the memoir, indicating through Shaw's own understanding of the political import of clothing her commitment, as well as Lucy's, to their New Life ideals. She relishes relating the story, as it marks a high point in her own political awareness, if not in communism. For Shaw, the laundry debacle represents anarcho-communist utopia as a material feminist space.

Despite men's failure to grasp the latter, Whiteway was resilient and survived, evolving for many years. In the foreword to *Whiteway*, one of the original deed holders, Joseph Burtt, writes of "an airy Cotswold upland, where frost did not chill by night nor the sun burn by day. I do not remember that rain ever fell in that sweet Arcady.... If our feet were down in the potato trenches, our heads were up with the stars. We felt we were gods."[109] In contrast, although Shaw describes Whiteway as a "little Utopia," she is very clear that its people are not "perfect beings." "But at any rate," she states, "we had a jolly good try, enough to show our sincerity.... What remained and still remains," she continues, "is the doctrine of 'bread labor.' Equally important is the idea of 'free land,' that the land should belong to the people. Free to be used by those who desire to earn their livelihood in that way."[110] It was deeply satisfying to Shaw that conventional clothing eventually caught up to the New Woman's rational dress; this was proof of her own insights into the link between outside and inside, comfort and freedom, ideas and materiality itself.

However, in many ways, Whiteway remains off the map of its local region. In one of the earliest incarnations of environmentalism, the Rothschild List, the Cornish coast was included as an area to be protected. The Cotswolds themselves are now designated a "national character area" by the National Trust. The region is known for its stunning views, its golden-colored stone-built homes, barns, and cathedrals, and intact nineteenth-century outfarms, farmhouses, and field barns.[111] But Whiteway is not found on the National Trust's map of the Cotswolds. Despite its nineteenth-century origins, it seems not to qualify as national "heritage," a memory to be preserved and carried forward into the future. In fact, neither Whiteway nor *Whiteway* are marked as a place or a book that matters, revealing that the practices of conservation and national memory keeping are highly selective. The gendered political ecology Shaw's memoir constructs, however, seems important to remember, as Sharon Butler and Peggy and Bert Bundy wrote in 1988. They object to the erasure of women's participation in utopian communities as "social-science fiction... not a realistic assessment of what was actually possible," and their essay on Shaw and George Egerton points out that Egerton's tale "The Regeneration of Two" may reference such historical projects that were "in the air" at the time.[112] However, Shaw's text still remains out of print and mainly out of our historical memory. To suppress it, as Greenway notes, is to do ideological work that curtails future thought. Such texts address a gendered perspective on how utopian thought might conceive resilient, convivial alternatives to consumer capitalism as well as the discourse of sustainability. At the same time, the text is, like Ellis's *Attainment,* an unflinching record of the gender practices of the New Life, revealing both its promise and the gaps in its practices. As the New Life stands to be embraced anew by ecocritics, it is salutary to note Shaw's perspective on Whiteway's "splendid attempt" at materializing utopia in an equitable fashion, in the tradition of the New Woman writer, bringing the future into the present.[113] This theme is explored as well in George Egerton's novella "The Regeneration of Two."

"Snow Everywhere!": The Heavens on Earth in "The Regeneration of Two"

The clothing of utopia, both in snow and in handmade, unconventional attire, enlists the power of the elements and the fabric of the earth to regenerate the sexes in George Egerton's "The Regeneration of Two" (1894). As the tale opens, the setting is a summer resort on a fjord carefully domesticated as a pleasure ground for middle-class travelers. In this world, "lilac" has ironically become merely the color of a stripe on a dress. The heroine, known only as Fruen or "Mistress," is initially represented as domesticated and detached, corseted, powdered, and parasoled. A not-quite-grieving widow, she seems listless and conventional, taking a great deal of time to dress herself, only to go seemingly nowhere. Traveling aimlessly around the fjord by steamship and strolling the resort's manicured paths, however, she encounters a man sleeping in the hot sun off the beaten path. She shields him with her parasol and watches him sleep for hours. When he wakes, the two talk; he, a poet philosopher, as it turns out, lays the ills of society at the feet of fashionable women such as herself. Launching a scalding Nietzschean critique of women in modernity, he accuses Fruen of coyness, of wearing a mask for men that obscures her own sexual desire and embodiment.[114] Angry at her inability to charm this man, the heroine is shaken enough by his critique to faint. When she awakens, she experiences an epiphany; she comes into full consciousness of what she realizes is a role she has been playing in a gendered masquerade.

By contrast, the arrival of winter—"Snow everywhere!"—in the second half of the tale naturalizes her "regeneration" during which she is materially transformed in a neo-pagan landscape, blessed by the gods themselves with snow.[115] In this landscape Fruen has even defied the local minister, "Herr Pastor," and transformed her fine estate into a dormitory for a "fallen" women's handicraft community. The grounds and outbuildings have become a farm and studio deeply engaged with working hands, weaving, planting, and caring for animals. The wintery setting elicits new understandings of materiality itself, as through snow "every outline is sharply defined" and newly exposed.[116] Out of summer's somnolence, the reader's senses are awakened; we see the nuances of the element of snow, its paradoxical weight and lightness, and especially its subtle spectrum of color: "silver-white, golden-white, white with a grey, and white with a green in it. The sea is frozen near the land into glass-grey ridges, and further out the waves wash over the serrated edges of the last freezing."[117] The snow is marked "in all directions" by "the prints of beasts, and the telltale impression of birds' claws" and yet "the heaviest animal goes with a padding step."[118] In this newly complex and lively landscape, the traffic between elemental, animal, and human bodies is suddenly legible, "marked." Fruen is marked too: "Her cheeks are glowing with cold and exercise. She looks a different being from the anemic woman of three summers ago." The Christmas tree itself bursts out of its Christian frame and becomes a kind of "a monster planted in a huge feeding-tub."[119] These images flesh out the colony's alternative ecology, solidifying its connections between people, the elements, and the winter village near Oslo, Norway. At times, the holiday season seems to circle around the body of Fruen, draped in red velvet and trimmed in sable.

Following late Victorian, protoeugenicist concerns with how the immersion in environments and climates could produce shifts in human bodies, this snowy climate in the second half of the story regenerates the heroine through its own icy power, schooling her in nonreproductive immanence and her own materiality and allowing her to move, like "the heaviest animal," with an ironic speed. As the tale progresses, charting Fruen's development of a community comprised of sexually experienced women rejected by the village, her search for bodily autonomy and pleasure takes place through her immersion in snow, her entanglement with matter, and her erosion of the division of human culture from the earth's wider materiality. In the women's community Fruen has formed, the land has become a vibrant partner. The women weave preindustrial patterns, using natural dyes and flax to produce linen garments based on antique designs. These garments visualize the ideals of transformation and regeneration. Moving heaven to earth, as snow materially does, Egerton establishes the dizzying possibility of myriad regenerations: in sex, labor, and the environment. As snow falls, it reminds the reader that, in action, "Nature does not love a ladder," but rather disrupts stable hierarchies. Snow melts or freezes, and brings the sky to earth.[120] It is always in a state of regeneration, of becoming something else, something new, and its presence primes the scene for unpredictable, perhaps limitless and pleasurable change.

In their new environment, "the women of Egerton's commune are big and strong and physically vibrant."[121] But the tale has grave difficulties in maintaining the material anarchy it explores through the element of snow. None of the women are bigger or stronger than Fruen. She describes them as "my" responsibility. This becomes useful when Fruen confronts outsiders like the local pastor. She tells him that "Man hasn't kept the race going, the burden of centuries has lain on the women . . . I know you don't agree, Herr Pastor, but we are doing very well; my colony of sinners almost pay for themselves."[122] Fruen's words, "my colony," aptly describe her relationship to the women in the community. She is, paradoxically, the "Snow Queen" placed in a commune.[123] At Christmas time, for example, in her crimson gown "she looks very big," and she reflects on how her dress is "spun and woven on her own place, and she is very proud of it; she has put sheep on a rocky bit of land and the wool is dyed after an old recipe."[124] The body that matters here is Fruen's as she celebrates her "present stage" of life.[125] "Her heart warms as she looks round her big kitchen, filled with people all dependent on her in some way."[126] Fruen has not just been embodied, but she has been embodied as a leader of poor women directed to create sustainable handicrafts in "her colony." As Egerton deploys elements of the fairy tale to aggrandize Fruen, she negotiates with time itself, making Fruen much larger than her historical moment, a hybrid of the visual power of the solitary New Woman figure and Hans Andersen's elemental queen. As Egerton does this, she fails to invoke the power of the collective, the alternative community, but she gains what José Muñoz describes as "the critical work that utopian thought does, in its most concise and lucid form, allow[ing] us to see different worlds and realities." The fairy-tale form may again queer straight time, bending back the power of "Herr Pastor" in Fruen's triumph over him. Such "conjured" realities may instruct the reader that "the 'here and now' is simply not enough."[127] This "big" world Egerton imagines makes other places, like reality, seem small.

The elevation of Fruen continues in the tale to its end, when, along with her poet lover with whom she is reconciled, she seems to remain in place as an anomalous leader of a cooperative. The other women in the community are clearly a backdrop, classed with the children, to this "big and bonny" leader and her man.[128] It is unclear if the women in the colony will remain celibate, but it is clear that Fruen will attain a free union with the poet philosopher she met in part 1. As he rhapsodizes about his love for her, "my queen, my love, my dear love!" the attraction between these two strapping individuals is, of course, big. So is their sin. In comparison to the "little sinners" she helps, Fruen's free union with the poet is premeditated, a deliberate political choice. The union is welcomed by a Greek chorus of snow, the little sinners, and laughter: "Snow falls softly and the darkness gathers, but inside the music of women's voices singing at their work and the patter of children's feet and cooing laughter fill the house in which love is making a carnival of roses."[129] Appearing in *Discords,* a collection that stresses the jarring dissonance and discontent women experience when their desires run aground on Victorian sexual and social mores, "The Regeneration of Two" resolves that central theme by marking off the utopian space and time of Fruen's commune, blessed by the regenerative power of snow.

Egerton's Fruen is extraordinary, as Lisa Hager notes, for her free expression of sexual desire, for establishing her agency in directing her own life, and for "coalition building among women," but the same is problematized as Egerton has difficulty "valuing each member's difference and agency."[130] While Fruen claims her pride in her sexual "right to dispose of myself as I will, to choose" her partner, a eugenicist ideal of fitness makes hers the body that matters. The colony will go on in an icy clime now lit from within by her "fire" that not only expresses sexual passion, but the disturbing purity or "clean[liness]" often visible in the New Woman.[131] At the end, Egerton links the "free man" and "free woman" with a fiery feminist purity discourse. She is "like a tall pillar of white flame ... glorious with a fire that is too clean, too strong for shyness." Egerton's fetishizing of Fruen's robust, purifying body enacts both the hypervisibility of the body of the New Woman and that figure's supervisory role over others. Ironically, the colony does belong to Fruen; it is placed on her own estate. It is not just people, then, who are regenerated and brought into the New Life in Egerton's tale, but it is places, particularly the countryside over the city, and in estates over more humble places. Modeling the estate as a place that might be opened to "the people" and regenerated for them is perhaps the central idea of early British and European conservation, and the latter is quite different from the demands of bread labor and the transformation of the land pursued in *Whiteway*. The preservation of the ideal space of the estate and the idealization of Fruen's body recalls the Countess of Warwick's equally feminist intention to place the "New Woman" on "Old Acres."[132] Warwick managed to transform Studley Castle into Studley College, a women's agricultural college. Thinking big certainly allowed such historical women a voice in England's earliest conservation model: the preservation of buildings and places. Ellis's cottage renovations, Shaw's memorializing of Whiteway on the Cotswolds, and Egerton's tale with its emphasis on a conservationist role for a grand woman and her grand estate suggest that the regenerative work of New Woman utopias may intersect with some of the earliest forms of English environmentalism: conservation and preservation, the listing of certain places that matter.

As Fruen goes back to the land and claims the land for the people, it becomes clear that she determines which land is to be claimed, for which people, and how. This is a strand of the New Woman's utopian literature and culture that continues in the eugenicist, conservationist work of Egerton's peer Sarah Grand. In her future thought, as well, it is possible to see great strides made toward rejecting capitalist industrialism and its abuse of people and the land. But, at the same time, women of a certain class determine what should be conserved and where. Their work anticipates a conservationist imperative that remains problematic in the twentieth century. The Rothschild List, for example, formed in 1916, worked to preserve places important to the wealthy, where rare plants and animals, distinctive landscapes, or structures of "historic interest" to them might be found.[133] To be sure, Egerton's transformation of her estate into a cooperative dormitory led by a robust "Mistress" aligns her work with this type of exclusive conservation, preserving not just buildings and grounds, but an elite woman's choice to interpret and shift their meaning for the future. Fruen's utopian colony aggrandizes the New Woman's privilege to identify heritage for others, a practice Sarah Grand also embraces in her long-neglected novel *Adnam's Orchard*.

"A Green Desert": Sarah Grand's *Adnam's Orchard*

In *Adnam's Orchard*, a new garden of Eden is created through intensive culture, under glass, only to be "smashed to atoms" by the novel's end.[134] The plot centers around Ursula Pratt, a prophetic visionary mother figure who supports her son Adnam in the formation of a utopian intentional community, an agrarian commune in which cooperative labor and technology reduce workloads and in which workers are uplifted. While the men practice intensive culture during the day, Ursula cultivates them in the evening, educating them in music and reading (of New Woman short fiction, of course). Grand's path to utopia is routed through a vision of local food production, cooperative labor (augmented as in Egerton by clearly authoritative and superior figures), and conservative feminist intervention that promotes iconic images of the New Woman. (This is clear in Ursula's selections of readings for the men, such as the tale of a self-sacrificing woman doctor in Elizabeth Stuart Phelps's "Zerviah Hope.") A maternalist, nationalist ecology updates the novel's biblical frame and its latent, familiar undercurrents of apocalypse and lost Edens. Grand's novel succeeds in avoiding the marriage plot and in constructing a vision of a nonreproductive, ecological future on the horizon. However, the novel's embrace of both ecoapocalypse and eugenics anticipates and aligns with problematic narrative patterns in early ecological thought.

Grand's market garden plot is quite clearly a metaphor for "weeding" out the "stock" unfit for England in the novel; how such plots inform early environmentalisms, however, is a question that remains to be asked.[135] Grand's interweaving of ecology with eugenics is present in the work of many early greens discussed here, in addition to Egerton. Nellie Shaw, for example, speaks of those who are "fit" or "unfit" to marry, and her vision of utopia, like Edith Ellis's and Egerton's, maps out a place for the ablest of bodies. William Morris's utopian thought also features beautiful, sound, white bodies in his postapocalyptic future. Ruskin's "Nature of the Gothic," moreover, explicitly complains that in England "we manufacture everything . . . except men; we blanch cotton, and strengthen steel, and refine sugar . . . but to brighten, to strengthen, to refine, or to form a single living spirit, never enters into their estimate of advantages."[136] Ruskin's process of "refining," "blanching," and "brightening" is an unsubtle code for whitening and cleansing. This ableist, raced ideology should, then, be understood as a problematic strand of early green culture that is clearly incorporated into twentieth-century environmentalism; the location of bodies that matter is linked to places that matter in the earliest examples of the latter.[137] In Grand's novel, Adnam and Ursula's superiority and the promise of their garden is heightened by the introduction of Adnam's degenerate half-brother, Seraph, and the high hopes of Grand's utopia are placed precariously on a garden of glass. Even as the novel idealizes fellowship, it slides toward a spectacular rupture. This threat anticipates similar imagery in the writings of Nazi R. Walter Darré, for example, who was to envision "the garden (i.e., society)," as a "breeding ground for the plants" which needs a fit gardener "to lift itself above the harsh rule of natural forces." The work of this gardener is to "provid[e] suitable conditions for growing." He will succeed by keeping harmful influences away, tending "what needs tending," and "ruthlessly eliminat[ing] the weeds which would deprive the better plants of nutrition, the air, light, sun."[138] This vigilant position against weeds naturalizes conflict and makes way for a spectacular, violent clash. Such plots exploding into the latter are firmly entrenched even now in ecoapocalyptic narratives.[139] Grand's novel then strikes several of the more disturbing keynotes of apocalyptic ecological plots in modernity.

The story begins when the unemployed, thoughtful, and restless Adnam, his father's second son, is nonetheless the first to see value in land neglected for years. At the opening of the novel, Adnam surveys an ironically fruitful and green orchard, which obscures fallen, uneaten apples rotting on the ground in "rank herbage." This waste seems invisible to his yeoman farmer father, who remains committed to monoculture. "There are bushels of apples there," Adnam protests. "They are good cider apples." The son ironically explains the value of apples to his father and asks for four acres of his own: "Let me have a free hand to make what profit I can out of them—for myself- I'd soon have a different tale to tell.[140] Adnam sources a book for all of his market garden plans, marking his knowledge as scientific, rational, and modern. What follows is the narrative of his slow rise as a market gardener over a period of three years (the typical time for the maturation of a market garden). In the meantime, Grand dramatizes and visualizes the land's neglect and the dense cultural web of unsustainable practices inherent to the agricultural depression otherwise invisible to the main characters and, arguably, the nation.[141]

The cause and cover of that crisis, Grand argues, lies in women's flower gardens. Enamored of fresh flowers and the "wild garden," women hold the nation in thrall to food importation.[142] Historically and in the novel, the countryside is dotted with roses climbing up dead trees or over decaying sheds or hedgerow, and garden plans include "the incorporation of bogs" and ditches into the landscape. These ornamental strategies were particularly useful "following the onset of the agricultural depression, which led to increases in fallow land, conversion of wheat fields to grazing or horticulture."[143] If "the wild garden offered the possibility of turning the detritus of the depression into an aesthetic statement that could erase the unsettling effects of the collapse of English farming,"[144] Grand exposes this ruse. She indicts the link between floral fashion and the depression, pointing out the invasion of weeds and the disappearance of local food production, jobs, and plants under the quaint, embowered countryside. Her critique wrings presence out of absence, making much of the biodiversity that has vanished despite the apparent greenness of the local landscape. Fields, she writes, are "gone out of cultivation, and left to lay themselves down in permanent pasture, of the scantiest, for want of tendance . . . hedgerows here, lovely in their wild luxuriance, were a delight to the eye, so long as the eye sought nothing but the beauty of wild luxuriance. . . . There was nothing for the hedgerows to shelter; the purpose of their existence had ceased to be."[145] The normalizing of the extinction of the countryside in the novel is maintained by traditional women who aestheticize that state of decay through the fetishizing of flowers.

Adnam's Orchard reveals that, while the English garden may "imagine the topography and scope of the nation and represent it back to itself," Victorian flower gardens—"exotic," wild, formal, or cottage style—pushed out food production and in fact were enabled by the outsourcing of the latter. The garden, Helmreich argues, "had particular resonance as a means of imagining nationness. An enclosed space devoted to cultivation and display of plants, the garden mirrored the notion of nationhood as a bounded territory designated for a particular set of peoples."[146] As a place apart from the labor of food production, the devotion of space to "landscape" or "scenery" indicated the nation's transcendence of necessity, need, or want. The landscape "garden provided a reassuring national image and became both a solace and a bulwark from the vagaries of modern capitalism and imperialism, acting as a physical and imaginative shield."[147] However, the same national emphasis on design and mere visual appear led to a loss of biodiverse, edible plants.

The prevalence of ornamental gardens also obscured the economic relationship between Britain's colonies and its food supply and the nation's intimate traffic with the latter through eating. Nothing more effectively concealed the inherent violence of the disappearance of food and medicinal plants from the countryside like the magnificent heritage gardens of England. One practice enabled another, as the land could only be turned over to ornamental gardening on such a large scale if it relied upon foreign food, outsourcing food production to the colonies via the latest industrialized large-scale farming and shipping methods, the train and the steamship. The domino effect of outsourcing, cheap imports, and bad English weather, as well as declines in land values, productivity, and agricultural employment, all added up to depopulate the countryside and leave rural England in decline.[148] At the same time, the Third Reform Act changed the political makeup of the countryside by extending votes to laborers who mainly lived in the city. The Redistribution Bill of 1885 had granted one man one vote regardless of property ownership; now urban representatives outnumbered rural representatives in Parliament for the first time in historical memory. Limitations were placed on the powers of the House of Lords in 1911, the year before Grand published *Adnam's Orchard*. The sense was that the country would be fundamentally threatened, in "danger of disappearing altogether."[149] Grand remakes the image of the national garden through intensive culture, nationalizing it, purifying it, and placing it in under glass in the novel, where it radiates possibility for the future.

By introducing a food-productive garden into the embowered territory of the English novel itself, Grand invades its potent floral discourse, developing a critique of floral cultural forms and their aesthetic expression in the novel, contrasting an excess of flowers to a scarcity of food plants. She links the New Woman's concept of the traditional woman's "parasitism" to their collection of flowers, which represents an unsustainable ecology that is held in place through the exclusion of local food production.[150] Her novel recalls her essay "The New Aspect of the Woman Question," in which she responds to Ruskin's "Of Queen's Gardens": "We have been reproached by Ruskin for shutting ourselves up behind park palings and garden walls, regardless of the waste world that moans in misery without, and that has been too much our attitude; but the day of our acquiescence is over."[151] Although much has been made of the New Woman's urbanity and her displacement from the garden through this landmark essay, it seems clear that the latter's domesticating discourse is shifted when the garden under discussion is edible. With the exception of pomegranates and artichokes, Ruskin's mythical garden in his famous essay grows only flowers. Ironically, a critique of such gendered floral gardens produces Grand's ecological thought.

Grand presents Ursula, then, as one of the novel's rare women with insight into the novel's troubling floral ecology. Other women "wax eloquent about the wild landscape."[152] They do not see that, despite its luxuriant appearance, the local ecology itself has lost biodiversity: "In front of each cottage, once bright with flowers and sweet herbs, and longer strips for vegetables at the back, now" all are replaced "with weeds."[153] The "dear weeds" beloved by women are better viewed with "the eye of the modern agriculturalist," who sees the landscape as "symptomatic of the threatened decay of a great nation.[154] Traditional women struggle with this view, particularly Ursula's friend the "little duchess" of Castlefield Saye, who is indicted in her flowery boudoir as a chief keeper of this kind of rustic decay. A foil to Ursula, the duchess is horrified by Adnam's market garden: "It looks like the abomination of desolation and mud pies! What a pity to spoil your picturesque old Orchard and that dear old field."[155] The duchess equates these "dreadful new ideas" with the apocalyptic end of civilization: "My husband says the country is full of such ideas and we shall all be ruined. Socialism, you know, and all that kind of thing."[156] The duchess's inability to recognize which practices are actually violent and destructive only makes Ursula's vision seem clearer.

Grand stages conversations between Ursula and the duchess to this end. When the duchess objects to the removal of "pretty" weeds for Adnam's garden, Ursula explains that "they were not wholesome, and there is no profit to be made out of weeds." Ursula's responses to her friend's conversation are often above her head but not the reader's. "My dear Ursula!" the duchess protests. "Surely you have not begun to talk like that! The duke says it is all materialism. People think of nothing but profit now and how to make things pay. You used to have ideals!" Ursula's response to the Duchess is disturbing: "I have still," Mrs. Pratt answered. "I want to see the weeds killed everywhere."[157] Ursula often gazes out her window, above and beyond her own flower garden, which she rarely acknowledges: "That at which she was gazing with strained attention was not in the atmosphere of the earth, but beyond, where the inner eye can see untrammeled by limits of time and space."[158] Ursula dreams of a utopia without weeds of the vegetative or human variety. She is a eugenicist for whom the garden is a step toward a wider transformative project of cleansing the land of the unfit.[159] Sweeping floral discourse aside allows for the rise of this visionary alternative New Woman figure and for the preservation of her home, the South Country.

Talia Schaffer has analyzed how, for Grand, "cleansing becomes the greatest women's virtue, not just for families, nor even just for England, but for the 'human household,' a semi-divine mission."[160] This enabled Grand to promote the language and iconography of a "utopian feminist vision of the future."[161] Developing mythic language, revising Biblical myth, and invoking mysticism "gives Grand license to create her own symbolic panoramas and sweeping mythic histories."[162] Bringing mythical language, iconography, and mysticism into the text through Ursula's visions, then, not only holds the novel's characters responsible for "vast declines and mythic changes," but in the intersection of this image with the novel's evocation of the garden and the fall, Grand's mythic New Woman strategy also makes way for regeneration through ecological apocalypse.[163]

The dazzling effect of the fragile glass in Adnam's market garden primes it for destruction despite the fascinating green modernity it promises. At "first glance" the garden offers a full dose of future shock. It

> gave a general effect of glass, glass everywhere, glass houses, glass frames, glass bells, the latter looking like a mushroom growth up-springing on broad beds with strange regularity. Bright rays and stars, dazzling to the eye, shot back all the colors of the rainbow from the glass to the sun, where it struck on a knob or a projecting angle. Under the glass, through a veil of mist, it was possible to see, by peering, that there was greenery. Acres of vegetables, not yet in season, stood there, warmly protected, and either ready or very soon to be ready for cutting.[164]

On the one hand, Adnam's dazzling glass garden "re-myths" Nature as "a fluid, changing, historicized, constructed concept" that can accommodate the presence of a utopian modernity without losing its right to exist as an "environmental" space.[165] In *Adnam's Orchard*, the practice of alternative agriculture itself is represented as a co-construction between the men of the novel, the elements of earth, air, fire, and water, and low-impact technologies like glass. The gardening techniques produce a viable, sustainable treatment of the land and are a step toward a more just—in this case, socialist—society. Ursula is the bearer and protector of this future; she holds "the map of Adnam's life spread out before her in her mind's eye." She tells her son, "Socialism, cooperation is the watchword of the future," speaking ironically from her throne-like abbess's chair.[166] "You inspire me," replies Adnam.[167] Amid this mother and child regeneration, however, remains the compelling darkness of Seraph, Adnam's half-brother. The text's acceleration toward an apocalyptic ending in which the garden is destroyed is signaled when Adnam presents Ursula with a "vivid scarlet" flower as a "red badge of courage!"[168] This gesture militarizes floral discourse and transforms Ursula into a martyred soldier as well as an aesthetic object. The flower remains a potent if not fatal entity. Soon after receiving it, Ursula dies, wounded by her own love for her husband when she is abruptly widowed. Ursula's radical social ecology also constructs Grand's conservative and maternalist cultural feminism, which now seems informed by blood and sacrifice. The destruction of Adnam's garden in turn violates the bond between blood and soil.

The attack on the garden proves the problem of human weeds and the justification for their eradication. As violence suddenly enters the novel, it provides a resolution to the representational challenge of creating a compelling narrative out of the slow, attritional violence of environmental degradation such as the agricultural depression and the slow solution of local gardening that has unfolded over many pages. Grand takes the opportunity to rewrite Genesis, laying the cause of Adam's expulsion from Eden upon a jealous brother. Although Rob Nixon notes that the realist novel increasingly reflects the pressure to incorporate a modern aesthetic that equates environmental damage with violent spectacle, particularly since 9/11, the end of *Adnam's Orchard* makes clear that this tendency has long been latent in ecological plots.[169] The conclusion savors the garden's destruction, picturing the orchard subjected to "to fiends bent on destruction. The bell glasses had been smashed to atoms and the young plants under them trampled in the dirt. The lids of the hot-beds had been pulled off and broken, and their contents torn up by the roots and scattered about. The hot-houses were a litter of broken glass and dying vegetable." The heat and the water in the garden have been turned off, subjecting the plants to flood rapidly turning to "slush and mud and cat-ice. Nothing that diabolical ingenuity could devise for the purpose had been neglected. Adnam's Orchard was a wreck."[170]

The "wreck" of the glass garden feeds an apocalyptic narrative that makes Adnam an exile from his own country. Grand's readers responded passionately to this scene: "Adnam's intensive culture experiment was so fascinating an experiment that we raged when all his work was trampled down."[171] As the novel ends, Adnam walks away from his local garden, "steer[ing] straight" for his future. From the "brow of the hill," he looks down on his home with its intricate classed patchwork of yeoman farms, tenant farms, derelict cottages, great houses, and "the Castle. . . . Henceforth, Adnam's Orchard was the

World."[172] This narrative trajectory is satisfying at the level of realist narrative, recalling of course the expulsion narrative of Genesis. Paradoxically, however, the violent spectacle of the garden's destruction, while evoking the "rage" of the reader in quite satisfying ways, also makes the environmental loss of the garden permanent and sends Adnam elsewhere. The slow story of ecological recovery drops off the page; that narrative too is "a wreck." The story of those workers who live with the local damage is not told. This narrative strategy anticipates twentieth-century environmentalist narratives that idealize lost Edens and represent environmental destruction as inevitable, the result of single catastrophic events, "explosive and instantaneous" as well as permanent.

Adnam's final vision of the landscape he is leaving behind reveals a deeply interconnected local ecology—yeoman farms, great forests, tenant's cottages, the Castle—in which access to resources is determined by class and "blood," and of which intersections he is an ideal product or "stock." As Adnam heads out to the world, the South Country takes on a new meaning as the lost garden. The South Country's loss is never seen as anything but a gain for the colonies, who "take" the money and the blood of the English. Grand has no awareness of the eventual environmental losses of those nations—deforestation, desertification—who become England's market garden. Her focus on local catastrophe and loss makes her part of a modern ecological consciousness that frames itself representationally through the local rather than through systematic links between nations. The contribution of the mythic New Woman Ursula to this troubling ecological narrative helps rethink the emergent role of women as consumers and producers of the discourse of heritage, localism, and bioregionalism in environmentalism in the early twentieth century and even today. Grand's alliance of intensive culture with eugenics, and her deployment of catastrophic or apocalyptic environmental narratives in which the underbred dark angels attack the future in the present, makes the practice of market gardening and intensive culture vulnerable to co-optation by nationalism. This was certainly to be the case only a few years after the publication of *Adnam's Orchard,* when women were asked to come back to the land to feed a hungry nation going to war, in part to keep its food importation routes open.

Utopian Dreams

New Woman utopian dreams and memories offer prescient, if sometimes disturbing, ecologies. Reading these from Ellis to Grand reveals glimpses of women's resistance to the Anthropocene, to straight, ecocidal time. At the same time, women's dreams of the future may create new concerns for people, places, and things.

The introduction to this chapter began by invoking Margaret Atwood's and Sarah Hall's more recent futurist interventions into myths of sustainability in *The Handmaid's Tale* and *Daughters of the North*. In Hall's novel, the protagonist is called "Sister." Her name addresses what is absent from the chilling neo-traditional familial relations of Atwood's novel, posing an alternative to the terrifying patriarchal nomenclature of Gilead. Both Hall's darkly utopian vision of sisterhood and Atwood's apocalyptic vision of complicit aunts, however, gain in complexity when placed on a timeline revised to include their historical and literary maternal antecedents. Reclaiming the meeting of New Woman literature and culture with the utopian thought of the New Life challenges the universalism of the prefix in the term "Anthropocene" and introduces a complex lineage of resisting sisterhood and their alternative visions of a green future. As flawed as it is exhilarating, this work has too long been out of print. The loneliness and isolation of Offred and Sister in the toxic dystopian times they seek to escape, the dehistoricized, anonymous documents from which their stories are told, and even the fragility of Sister's treasured collection of old newspaper clippings that enables her search for the alternative, sustainable women's community of Carhullan, resonate with the historical and contemporary need of women to locate their own literature and culture of resistance, the situated knowledges, and indeed, the ecologies, shaped by their dissenting voices. These, like Nellie Shaw's examination of what comes out in the wash of communism or Edith Ellis's ecstatic queer calendar and cottage experiment, are materials through which even now we might create a more equitable, inclusive environmentalism. Their feminist, land-based activism provides an immediate historical context for Egerton's and Grand's utopias. These, in turn, provide timely food for thought in contemplating our own ecological strategies, which may foster resistance or be easily co-opted by the interests of the nation. This was the case, in fact, during the formation of the Women's Land Army during the Great War.[173] During that national crisis, as well, women created an invaluable record that resonates today.

4

"GOD SPEED THE PLOUGH AND THE WOMAN WHO DRIVES IT"

Ecologies of the Great War

Early green women like Frances Greville Warwick, founder of Studley College, hoped that the Great War might "soun[d] the death-knell of the old industrial, social, and political conditions," both reviving the growing of local food and making the latter a permanent career option for women.[1] She was joined by activists like the Pankhursts, who hoped that women's work on the land would finally prove their worthiness for the vote.[2] These women followed Olive Schreiner's imperative that women "take all labor for our province!" whether "intellectual or physical toil."[3] Now, newly credentialed women such as agriculturalist and author Louisa Wilkins (1873–1929), among the first to complete an agricultural course at Newnham College at the turn of the century, and Meriel Talbot (1866–1956), who was to become the director of the first Women's Land Army, supported the idea that the nation's urgent wartime need for food might be an ideal opportunity to prove that agriculture was a field suited to educated women in the long term.[4] Much was at stake in the historic venture of women's wartime land work during the Great War. However, this chapter proposes that the literature and culture of wartime women land workers is also timely. It reveals how interests usually seen as separate—for example, gender roles and the practice of monoculture—are entangled in national ecologies through basic material practices such as eating.[5]

For individual women like artist and militant suffragette Olive Hockin, the physical challenge of landwork was an opportunity to prove the New Woman's creed that women could do "any mortal thing" they wished with their minds and bodies.[6] The official rhetoric of the Land Army and Women's Auxiliary Corps also encouraged women to prove themselves: "You have been given exceptional opportunities of making a career for yourself and at the same time doing your duty to the country which has reared and protected you. Make the most of your chances and be worthy of the trust that has been placed in you."[7] The literature and culture of wartime landworkers, however, tells its own story of human and other bodies as proving ground. For Olive Hockin or Rose Macaulay, as well as those often anonymous women who published their letters, cartoons, and essays in the Land Army's official journal, the *Landswoman,* learning where food comes from was a transformative experience that revealed the complex entanglement of classed and gendered labor practices long assumed to be natural or wholesome.

Often working against the grain of Romantic ecologies that find a path to spiritual awakening through contact with the environment, like soldiers in the trenches by the end of their service, many women emerge from their own service with questions. Some are disenchanted with the division of labor on the farm or with ploughing and weeding under harsh conditions. The ecological and labor issues of the latter are exposed, in fact, by the nation's ill-advised attempt at a precipitate return to growing corn and meat and the acceleration of planting and the disruption of the normal rotation of fields to do so. In response to what is often described as an attack on pasture land turned to tillage, soil grew weedy and crops failed to thrive. The result is the land worker's protomodernist suspicion of authority and even of the sustainability of a food like bread itself. Often, their work dismantles official national discourses along with the nationalist myth of England as a green and pleasant land. For them, tracing where one's breakfast comes from becomes a deeply affective encounter with plants, animals, and people, challenging vertical divisions between people and the land, culture and nature, as well as false binaries between city and country, and classed and gendered divisions of labor.

When scholars study the Land Army, it is often to read the extent to which it succeeded or failed as a gender or an agricultural experiment, paying particular attention to the degree to which official propaganda casts women in "nostalgic" cultural roles, growing food on the land as traditional nurturers or carers.[8] Little attention has been paid, however, to two issues. The first is the link between the Land Army and the Land Question as engaged by those women who had historically objected to food importation on political grounds and sought to professionalize women on the land as part of the New Woman's wider project of reclaiming labor. The second area of neglect is the study of the often rigorously analytic texts produced by women wartime land workers themselves. Written in the New Woman tradition of self-analysis and experimentation, these texts are often comprehensive unpackings of timely issues such as what are now called "food miles," the politics of importation, the unexpected agency and vibrancy of the land, and its gendered, ethical use. Tasking women with growing food for fighting men is clearly problematic. Women like Rose Macaulay question if they are actually holding their politically preferred status as "noncombatants" when they feed the war effort. Gendering any environmental work—such as recycling today, for example—is equally problematic, as it unfairly burdens women, provides another venue for their domestication, privatizes environmental action, and releases men and the public sector from that important work. What must be parsed from this critique, however, is the implication that *growing food itself* is essentially a nostalgic or degrading practice. Such a belief reifies nineteenth-century nationalist liberalism through which nations seek to transcend their material needs, shunting them onto other, often more vulnerable nations and peoples who must bear the brunt of agriculture's human and ecological costs.[9] The latter are significant, historically producing deforestation, drought, and famine, as well as more pollution than any other industry.[10] What remains to be done is to claim the literature and culture of women landworkers themselves as an important body of work that contains often powerful analyses of food cultivation and monoculture, as well as class and gender on the farm.

The often urban and inexperienced women land workers of the Great War, in fact, are powerful analysts of the cost of sustaining traditional practices in their own country. Their records of the realities of wartime landwork and particularly their sometimes agonizing, detailed accounts of the nation's extraordinarily misguided attempt to restart traditional monoculture on long-fallow land reveal a powerful critique of monoculture and its attendant ethical positions, such as the familiar environmental "stewardship" ethic.[11] The latter was to have a heavy hand in their project, leading to land seizure policies enabled by the Defence of the Realm Act (DORA).[12] Ironically, while previously the British farmer had received no support from the government under laissez-faire liberal policies, now the government was empowered to take over land deemed badly farmed. Under the Cultivation of Lands Order, War Agricultural Executive Committees were formed in each county to help increase food production in Britain. They enforced the Orders to "Plough up Pasture for Crop Planting," which aimed to bring fields used for grazing into arable production and to add two million additional acres of ploughed grazing land.[13] The campaign was not as successful as hoped; farmers had no choice of which fields to plough for a particular crop, and many farmers only had experience with livestock, as wheat had so long been displaced by cheap imports. Moreover, even for those farmers who were still in business growing vegetables, the push to accelerate rotation of their fields to grow more food more quickly often resulted in weeds as "big and flourishing" as the vegetables.[14]

The gap between popular and much-studied propaganda such as the famous poster that blares "God Speed the Plough and the Woman who Drives It" and the lived experience of women attempting to restart monoculture during the war, providing the bread and meat considered essential to fighting men, is a wide one. Instead, landswomen often less satisfied and newly suspicious of authority and official national discourses, dismantle the nationalist myth of England as an abundant and tranquil farm. This is clear in many textual examples to be studied here, from Olive Hockin's often brutal *Two Girls on the Land: War-Time on a Dartmoor Farm* (1918) to the deep cynicism of Rose Macaulay's "On the Land" poems, which include ironic titles such as "Picnic—1917," "Spreading Manure," and "Burning Twitch." Such work indicates many of the themes of New Woman ecologies already encountered, the shock of the often urban woman's encounter with an unexpectedly vibrant and active landscape with its own type of agency, the leveling of human superiority to the more-than-human world, and the critique of outmoded ecologies clustered around notions of gendered floral culture, extinction, and degeneration.

Like any army, the women involved in wartime land work were a diverse group, including recent college graduates like the seventeen-year-old Mary Lees, the suffragette, convicted arsonist, and artist Olive Hockin (1881–1936), and the soon-to-be poet and novelist Rose Macaulay (1881–1958). However, their subject position in the Land Army often dovetails with the image and the promise of the New Woman in strategic ways. Women's lived experience on the land was at once shaped as a "modern," military, and scientific occupation rather than a natural link.[15] The propaganda and the rhetoric of the Women's Land Army attempted to "broadcast to the public as an organization that offered women new opportunities during wartime," and it co-opted the visual culture of the New Woman. At the same time, officials shied away from "disrupting accepted pre-war gender codes." This obscured "the politically charged atmosphere within which the Women's Land Army was formed and operated."[16] Promised a modern approach to agriculture, recruits were in actuality to engage in the backbreaking work of monoculture, ploughing with outmoded tools in infertile, weedy files and with often aging or badly conditioned horses rejected for military service.[17]

While women like Cockerell and Nussey in cutting-edge programs had struggled not just to enter into the agricultural labor market, but to reshape labor so that it might be more cooperative, less competitive, more rational, and less brutally punishing to both human and more-than-human bodies, the Land Army returned to the traditional work practices associated with monoculture. Landworkers' representations of their work reveal what food scholars now note: food cultivation and consumption is an intimate engagement with "a host of social relationships such as those with the people who grow, harvest, or trade" food and also with "Nature, not in the abstract but with particular nonhuman others, things, and individual places."[18] In critiquing the lack of "cooperation among Dartmoor farmers," writers like Hockin make it clear that the result of their gender experiment is a sometimes shocking, fundamentally ecological view of the interdependence of life.[19] The official attempt to transcend the conditions of life had caused the egregiously late formation of the Land Army. The latter was not created, as Vita Sackville-West notes, until three years into the Great War, when "there was only about three weeks' food supply in the whole country."[20] Only then did the minister of agriculture decide to organize existing agricultural units of the Women's War Agricultural Committees (WWAC) into a nationally organized, quasi-military effort, a "Land Army."[21]

Land Army writing often shares insights into the land as a contested site where official ecologies and practices are analyzed, dismantled, and reshaped to incorporate the women's often startling realizations into tasks they themselves once assumed to be quite benevolent. Indeed, reading their work lends a disturbing and ironic new resonance to the time-honored image of the Great War as a "bomb in a garden party."[22] Landworkers come to critique an assumption that is clear in Sarah Grand's prewar novel, *Adnam's Orchard*: that a garden nation with nothing to eat conceals its capacity for aggression within itself—a hungry nation is an aggressor in the wider world, demanding its needs be met elsewhere. Landswomen experience this otherwise hidden relationship between food importation and global power as startling new knowledge, even as they come to understand the vulnerabilities of local ecologies and the vitality of climate, soil, and even, or especially, plants and animals.[23]

From the journal the *Landswoman* to the memoirs of Daisy Warwick and Olive Hockin, as well as to the bitterly eloquent collection of poems "On the Land" by former landworker Rose Macaulay, the literature and culture of wartime landworkers records the material meeting ground of perception and reality, over which landswomen cross into a political and critical consciousness one might describe as modernist in its alienation from official narratives, projects, and even forms of expression. Reading the Land Army's experience into and beyond its meaning for gender difference drives toward a recognition of the important role land use and food production played in the Great War as a catalyst for women's modernity. After the war, women, once promised permanent professional work in agriculture, were pushed off the land, and the country returned them to a dominant narrative of leisure or reproduction during the Second World War. As the nation reembraced the mass importation of food at nearly the same level it had pursued before the war, it defined the modern through the dual domestication of women and the erasure of the visible labor of food production. This is epitomized in shifts in propaganda and official discourse during the Second World War, when the Land Army's journal is renamed from the *Landswoman* to the *Land Girl*. Replacing masculine New Woman fashion imagery with images of bleached-blonde girls in sweaters driving tractors is also a powerful indicator that land work was no longer to be a career for women, but rather a temporary occupation conducted by young girls. The work was reaching its expiration date upon the end of the war, when women were to be returned to their primary work of reproduction. What remains vibrant, however, is the substantial body of work women landworkers produced that speaks even now to local and global food justice.

Gendering the Land Question

Today, national narratives forming around the memory of the Land Army often bury the issue of England's longtime importation of food, which created the wartime food crisis that necessitated the formation of the Land Army. It is well known that, during the war, "female political and social activists such as Harriot Stanton Blatch, Countess Frances Evelyn 'Daisy' Maynard Warwick, and Lady Gertrude Denman utilized agricultural labor shortages in Great Britain and the United States to improve the position of women in labor and to increase the momentum of the global female suffrage movements."[24] However, the agricultural labor shortage did not arise with the war but rather with the agricultural depression underway since the late 1870s. Thus the official language of the war's replacement of male agricultural workers with women is misleading. In most cases, men were long gone from the job of feeding England locally. Instead, tying the Land Army to both the suffrage movement and the Land Question and early green women's cultures reveals a more complex gendered environmental story. In England, women like the Countess of Warwick, founder of Studley College, had long insisted with William Morris and other socialists that "England should feed her own people," and that cultivation should be modernized, made cooperative, and gender equitable. Reviewing their historical positions on monoculture, women's labor, and land use reveals that the story of the Land Army involves a struggle not just over who was to farm England, but of *how* England was to be farmed.

Long before the war, Warwick had linked "the New Woman and Old Acres" in a productive new project that had little to do with traditional concepts of

Nature or farm labor. In her advocacy for women on the land, she encouraged cooperative, "light" labor strategies such as those demonstrated by Cockerell and Nussey in their market garden. Intensive culture was to be used as a strategy for addressing women's physical strength and the nation's poor soil still recovering from the advance of monoculture during the Age of Improvement. Her work linked both the Land and the Woman Questions in a mutually beneficial way. Warwick founded her agricultural programs to advocate for women's role in local food production as part of a strategy for a more class- and gender-equitable society from the late 1890s. Quoting Morris frequently in her book *A Woman and the War*, Warwick argues that food should be considered national wealth and the absence of a domestic harvest was not only immoral, but it made an ecologically vulnerable nation more warlike: "At present time," she writes, we "only produce about twenty-percent of the food we eat. For the rest, we depend upon our mercantile marine and our power to hold, not only the seas, but the skies above."[25] Warwick links the nation's faltering agrarian tradition to global industrialized capitalism, pointing out the intertwining of the nation's ecological vulnerability and military aggression; she uses the nation's food problem to recruit female defenders of local food as she critiques the inevitably aggressive posture in the world that England's food importation produced. The importation statistics themselves, she argues, "should suffice to settle the career of many a sturdy country-loving English girl" for the long haul.[26] With other revolutionaries like Peter Kropotkin, she rethought the benefits of the plough and monoculture, training students in alternative agriculture. This, she argued, was more ethical, equitable and peaceful, far less damaging to the soil and the human body, and allowed, as Grand points out, time for leisure, creativity, and culture.

But, as Warwick wrote, "war is ever fatal to ideas."[27] The agricultural innovations she had favored were overturned by the national narrative that bread and meat were crucial to feeding soldiers. The propaganda images such as the famous "God Speed the Plough and the Woman who Drives It" poster (see fig. 10) co-opt the image of the progressive New Woman alternative agriculturalist for the traditional practice of monoculture, once rejected by the same group. Comparing the *Punch* cartoon of the early green woman alternative agriculturalist to the "God Speed the Plough" recruitment poster makes clear that the former's alterity is carried over into the latter, where it is resituated in traditionally masculine labor: the short-haired woman wears boots, breeches, a fitted waistcoat, and a hat, but she is placed in a field, ploughing effortlessly into the horizon. Her masculinist appearance and domination of the land through the cutting plough, her forward motion, her relationship to

Figure 10. Women's Land Army recruitment poster, 1917. (© Imperial War Museums [Art. IWM PST 5996])

the rising sun, and even the poster's color, yellow, which was associated with fashionable New Woman culture, all construct her as a New Woman while the text and plough sacralize and ground her work in the most traditional form of masculinist labor. But, unlike the *Punch* cartoon of the alternative agriculturalist deliberately thwarting the stable order of things—soil itself—the landswoman's image makes soil something to be penetrated and returns agriculture firmly to the countryside, animals to subordination, and the New Women to a national ecology in which they only *temporarily* occupy a male role. The "God Speed the Plough" poster works to erode historical memory of the labor of the alternative agriculturalist, replacing it with an image of a New Woman performing not the modern work of "intensive culture" but the traditional sacred man's task of monoculture in a traditional field.

Warwick complained that "men whose minds were being turned slowly and reluctantly to questions they had been educated to ignore are now" merely "concerned with two problems—winning the war and making good the injuries it has entailed . . . there is a danger that the social movements, slow in times of prosperity, will halt in the times to come."[28] The question is, Warwick wrote, "What has woman gained" through her labor during the war, "and what has she lost?" It was too possible her service would be a zero-sum game. Warwick hoped to achieve a revolution that would transform gender roles as well as class stratification and labor practices. She urges readers to take up this plan in her memoir, where she describes British men and women equitably as "comrades." That word, she argues, "sum[s] up the best of the ideals of the human race."[29] She identifies with other "publicists of the social reform movement," who anticipated that the war itself would radicalize working people and result in a full-scale revolution.[30] "Whether we win or lose I see civil unrest inevitable, for this war has sounded the death-knell of the old industrial, social, and political conditions."[31] "Let us remember," she adds, "that the working classes that come back from war will have forgotten what fear means."[32] Warwick anticipates that after the war people will rise up against brutal labor practices, including those of traditional agriculture.

Describing the war as "man-made," Warwick holds out hope that women could intervene in its progress and perhaps make gains despite the risks they would face in allying themselves politically with men.[33] Her cultural feminism advocated the superiority of middle-class white women: "Those who rule over Europe and, being unable to settle their differences, sent millions of men, who have no quarrel, to deface the earth and slaughter one another, are morally responsible for every change in the normal life of mankind. Those who replenish the earth are better than those who destroy it. . . . War is a monstrous immorality that seeks to destroy the world."³⁴ Warwick's position is clear, and this is perhaps one reason why she was not given more to do—other than provide intensive training—in the hierarchy of the Land Army.

Other women, such as Louisa Wilkins, author of the Board of Agriculture's blue book on *Agricultural Education for Women* (1915), express little antiwar sentiment and remain strongly cathected to the idea that the war will produce long-term work for women. They write of the effect of the war on "educated" women, arguing that now "many more educated women than usual are faced with the prospect of having to earn their own living, and among them an appreciable number would prefer an outdoor life to a sedentary occupation in town. This question" of women's agricultural work "is a topical one, and will also remain permanent for this generation. . . . However acute this question may become" during the war, "it is undoubtedly a question for all time."³⁵ Wilkins was a leading figure in the Agricultural Organization Society. She used her "intellectual strength and organizing ability" to bring women like Meriel Talbot and Frances Wilkinson, the principal of Studley's rival, Swanley College, in to form the Land Army. From the beginning, these elite women set out to use "educated women" to act as leaders and trendsetters, problematically ignoring local women already engaged in agriculture, and subordinating the knowledge of women without formal schooling. "One educated woman," wrote Louisa Wilkins, "by her mere example and encouragement and powers of organization" might popularize and normalize the presence of women in agriculture.³⁶ Not only would she attract other educated women, but she would intervene in an ever-widening opposition between men's and women's work on the farm, attracting "village women" currently reluctant to do agricultural work.³⁷ Such classed New Woman strategies that privileged "educated" women over "village women" often led to intense conflict on farms between women and expressed an ironically missed opportunity to foster cooperation between classes of women who often found themselves alienated from each other. The organizers of the Land Army actively deployed such elitist strategies to communicate their hopes for a new incarnation of women on the land. But this new vision was exclusive, and it was aimed at the women who mattered the most to the Land Army's leaders, women like themselves. Propaganda images portrayed women on the land with bobbed hair and often stressed their mobility and class elitism through the presence of horses and equestrian attire, as is clear in cover art for the *Landswoman*. They staged mobile public spectacles and marches that included landswomen in uniform holding banners, leading livestock, carrying agricultural tools, driving tractors, mixing the discourse of Merry England with the visual shock of the New Woman.

However, unlike the urban settings in which many other "female firsts" in shipbuilding, mechanics, construction, roofing, bricklaying, and architecture took place, the agricultural settings in which Landswomen were pictured were powerfully associated with an abstract, gendered Nature itself. The New Woman's presence on the land signaled an entry into a postpastoral consciousness that countered women's continuity with a mythical green and pleasant land. If, despite their urbanity, education, and femininity, women fulfilled their orders, they defied national myths of a pastoral nation grounded in a tradition of a feminized, objectified land. They could appear instead as an "instrument of self-determination," which made women "appear somehow wrong—mannish, cheap, or superficial."[38] In this way, they opposed the conservative gender values that often went hand in hand with "Merry England," in which women were objectified as objects of desire, pursuing "light," attractive tasks such as sheepherding and dairying, both of which fostered the notion of their essentially nurturing, reproductive immanence. Such images, like those published of streetcar conductors in urban settings, "were less a record of exceptional individuals than a sign of the abnormality of the times."[39] Clearly, women in the Land Army intensely challenged the naturalizing of passive womanhood against the backdrop of their freshly ploughed fields. Scholars should remember, however, that landswomen were not defined exclusively through their founders' propaganda, official rhetoric, and their official portraits. Landswomen also mediated their experiences and images themselves by participating in the print culture of the Great War, particularly in the Land Army's own journal, the *Landswoman*.

In Their Own Words in the *Landswoman* Journal

The *Landswoman*, edited by Meriel Talbot, was the official monthly magazine of the Women's Land Army and the Women's Institutes. In her foreword to readers in the first issue, Talbot explains that the journal exists so that the Land Army "might have news of one another, and have a paper of their own through which to make known their ideas and their needs."[40] The Queen, Talbot affirms, gives "our magazine" a "right royal sendoff." Under a portrait of herself in an elegant hat and lace blouse, Talbot is also magisterial, hoping nonetheless that "it is as friends . . . we shall often meet one another in these pages and as comrades in the service of our country."[41] As Talbot states in her dedicatory to the first issue, the *Landswoman* exists to provide news relevant to Landswomen, but also to provide a space for them to publish their own images, reflections, poetry, letters, and articles. The Landswomen's own contributions appear side by side with a healthy dose of advertisements, official information, and instructive reading. The *Landswoman* was welcomed by workers on remote farms, like Olive Hockin, who saw it as a sign that the "present organization of women's agricultural work" would legitimate and protect the rights of landswomen, enabling the organization to "keep in touch" with otherwise isolated women.[42]

Like other periodicals of the time, the *Landswoman* had its own cartoonist, who signed her work "Bunty" and shared clever insights into agricultural work, propaganda images, and the gap between the representation and appearance of land workers in mainstream culture and the realities of their daily experiences. In Bunty's cartoons, women new to agriculture literally muck in to raise plants and animals, often working in clearly hostile conditions under the eyes of disapproving farmers, a group significantly absent from propaganda. Her cartoons often oppose those images and reflect instead upon the absurdities of the militarization of agriculture and the intensity of women's new intimacy with animals, soil, and food cultivation. They often display a significant critique of those in charge. This is clear in Bunty's October 1918 "The Recruiter," in which "The Recruiter" leans over a ledge, thrusts out her arm, and grasps at her audience.[43] She is dressed in official Land Army kit: belted overcoat, smock, hat, armband, breeches, short boots, and cockade. (The fact that few women actually received this uniform is made clear in the photographs in the journal of women actually working on farms.) The recruiter's face is a series of harsh angles; in this and other images, Bunty plays upon the aggressiveness of recruiting and the gap between the agency of this tailored, militarized image and the bodies of land workers. The cartoon is ironic because it defies the earnestness of the journal's official dedication and its editorial selections of pastoral short fiction, as well as the inclusion of multiple advertisements through which women might "erase" the presence of the land on their skin. Despite its official narrative, the material within the journal often provides a forum through which the recruiters' invisible audience, the landswomen themselves, might speak back to their leaders, recording their visceral experiences on the land, the disorienting new balance of power they encounter, and the shock of the vibrancy of the land itself.

Visualizing this experience, Bunty's cartoons are often bursting with the clashing of a variety of experiences and life forces. In the *Landswoman* of September 1919, for example, "Imagination" and "Realization" juxtapose two images.[44] The first shows a pristine landswoman carrying a bucket, her hat at a jaunty angle. She wears a kerchief, a filmy blouse, and lovely boots (as landswomen often do in propaganda images). However, on the page, "Imagination" looks not at the viewer, but at the second image of herself, entitled "Realization." As their eyes meet, "Realization"—who is wearing breeches, heavy clogs, an overcoat, and a well-secured hat and holding a pitchfork—stands on thick clots of earth in a ploughed field. With her back to the viewer, "Realization" reaches out to "Imagination" with an outstretched hand, arresting her swift movement. The cartoon suggests the extent to which work on the land transforms not just women's appearance, but their state of mind. Bunty's cartoon points out the landswoman's self-awareness, her ability to discern her own perception of herself from her representation in propaganda images. The cartoon references women's abilities both to imagine and realize their new roles on the land, detailing the transformative effect of their material encounter and the complex effect of the latter upon subjectivity. Although the image is triumphant, it is importantly not transcendent; rather than deepening the wide gap between agentic natures, it shrinks them. Indeed, in the cartoon the gap between nature and culture is narrowed. In covering the woman in mud and linking that experience to her realization of her subjectivity, the cartoon comically undoes the mind/body dualism, the masculinist wish to master the earth. The cartoon's opening of the landswoman's body to that element, its marking her with soil, creates a modernity that allows both for the soil's power and human vulnerability.[45]

As they explore the gap between propaganda and reality in cartoons, drawings, letters, poems, and photos in the *Landswoman*, recruits protest the idea of transcendence of the material, need, or necessity. They show a marked cynicism both about the idealized, clean women presented in the journal's advertisements for "vanishing" creams to remove the marks of the soil and the elements as well as their project's official task. The shifting ecology of the landswoman, in which she comes to know the vitality of matter itself, is wittily portrayed by another cartoon entitled "Molly on the Land/The Land on Molly," which also explores what is beyond the pale of the propaganda posters, honoring the realities of the common propaganda phrase "women on the land."[46] In the first cell Molly heads off to work, accompanied by the rising sun and singing birds. The silhouette of her body is trim and lithe. Ready to work, she carries her shovel in front of her and is walking on fresh, clean grass. In the second image, "The Land on Molly," she is walking home in the rain and mud. Her boots now seem more like clogs, clotted with mud working its way up her breeches and uniform. Her shovel is also marked with mud and is held behind her while her body is bent against a blowing wind. Now wearing the land itself, the landswoman is marked by her work. While recruitment posters featured clean women who worked "on the land," this image—like "Imagination and Realization"—represents the land's capacity to mark the landswomen as workers. Even the name "Molly" contrasts with idealized, classed representations of landworker outfits, which are advertised in the journal with glamorous, elegant names: "the Banbury, Doris, Beatrice, or Thelma."[47] Most of all, "Molly on the Land/the Land on Molly" reverses the agency of the popular phrase, which builds subjectivity through the woman's power to draw resources from the land while she herself remains transcendent, untouched and unchanged by her contact with it. "The Land on Molly" has the opposite effect. She is immanent rather than transcendent, not just covered in mud but sinking into it as she walks. Throughout the journal, such representations of landswomen's lived experiences resist official or commercial rhetoric, which seeks to erase both the mark of the land upon women and their vulnerability to the vibrant will of more-than-human nature.

Clearly, these cartoons oppose the visual rhetoric of erasure promoted by the advertisements that they often abut in the journal. Oatine Face Cream, for example, shouts at "Outdoor Workers," urging them to save themselves from "all weather conditions. Rain and wind and hard out-door work makes the skin rough and sore unless Oatine is used regularly; it possesses special properties which keep the complexion and hands clear, soft, and velvety.... USE IT AND PROVE IT."[48] A sparkling white-complected Landworker smiles in the ad. That cult of the erasure of visible knowledge and experience of the land is directly confronted by landswomen's own representations of themselves, in which visible signs of their experience—dirt, "weathering"—signifies their new knowledge of food production and women's labor. Submitting their own writing to the journal in forums such as "Farm Sketches," landswomen reflect upon the pains and pleasures of their own experience in their own words. Appearing right above the cartoon "Molly on the Land," a landswoman writes in to protest the notion that she and her comrades *should* be unmarked by their experience, for

> we are the world's workers. We feel this fact is not sufficiently recognized.... Firstly we are told our clothes are out of date; secondly we do our hair all wrong; thirdly our hands and noses are sunburnt and freckled—this is an unforgiveable sin. In vain we plead that were it not for us the meat they chew and the bread they eat would not be there ... we return to our bullocks and fields very grieved and fretfully injured—that's why we're writing this—we shall feel much better afterwards! Technically, we are known as 'Women on the land'—it always sounds as if everyone else lived in the air.[49]

Clearly, this landswoman resists being turned into a passive subject for the visual pleasure of men.[50] She struggles to make the nation recognize her as "the world's worker," who is marked and ennobled by her labor.

Her experience, moreover, has led her to understand that labor was previously made invisible in her culture. She shares the remarkable insight that *everyone* lives "on the land," no one transcends it, none of us lives "in the air," and all of us have to eat. This landswoman articulates a basic ecological principle: the recognition of human entanglement with, not transcendence of, the material world. She recognizes the role leisured women in "the pages of *The Tatler*" play in preserving the national fantasy of transcendence of the material world, "as if everyone else lived in the air." As Alaimo and Plumwood write, critiques of the false myth of material transcendence and the recognition of human vulnerability are deeply feminist, fundamental to a critical environmental feminism that dismantles enlightenment dichotomies that dictate that the mind transcends the body, that mental labor is superior to physical, and that only women are immanent, attached to the ground and their bodies.[51] The landswoman above seeks to resist the impulse of modernity to normalize that transcendence as she resists the pressure to remove the marks of her interdependence with the land from her appearance.

While official photographs of landswomen attempt to preserve an essentialist fantasy, obscuring landswomen's labor and reducing them to objects or models, landswomen themselves often protest this objectification. To reclaim themselves as landswomen is to point out a complex, gendered position held by recruits in the Land Army, including the position that the nation ought to feed itself rather than make a bid for a modernity that outsources food cultivation, transcending a basic human necessity. The disruption of the traditional identities "man," "woman," farmer, countryman, and their reduction to "landworker" avoids the fact that food "carries in it, and into us when we eat it, a host of social relationships such as those with the people who grow, harvest or trade" it "and also with Nature not in the abstract but with particular nonhuman others, things, and individual places."[52] When landswomen assume that land is merely a "resource" to be "instrumentalized" by humans for their needs, the land speaks back to them—loudly.[53] Debra Cohen has written that women's representations of the "Homefront" in the Great War remap the terrain of war itself; home front and front line are intimately connected and even symbiotic in women's wartime writings.[54] However, landswoman writings show that this symbiosis does not occur just at the level of discourse or narrative or even in the making of instruments of death, but it occurs, ironically, materially, in the making and consumption of the stuff of life itself, the growing of food—something landworker Olive Hockin learned on the moors of Dartmoor.

"Any Mortal Thing"

In providing women landworkers with a forum for their own often unvarnished discussions, the *Landswoman* shapes the culture of the Land Army, providing them with a common vocabulary, literature, art, reviews of relevant books, news, and images, some of which challenge, most of which promote and naturalize, a feminized, patriotic women's agricultural culture. The journal often shares pastoral poetry, including, for example, Alice Meynell's "The Shepherdess," the preferred reading of which, in the context of the journal, promotes the metaphor of women as nurturers and carers, extending the notion of the "Angel in the House" to women on the land.[55] But by August 1918, the editors also have in hand literary texts written by landswomen themselves. Meriel Talbot, the journal's editor, comments that the land has "made poets of all of you girls," and the *Landswoman* acts as a conduit for such texts, particularly the "happy" ones.[56] However, happy literary responses were not always available and, in their absence, the journal also published reviews of more problematic texts, personal memoirs, like Olive Hockin's *Two Girls on the Land: War-Time on a Dartmoor Farm* (1918). Meriel Talbot's conflicted review of this "freshly written" and "absorbingly interesting" memoir is a valuable representation of the gap between official and individual landswomen's voices. As Talbot acknowledges, Hockin's memoir provides a compelling and often devastating analysis of the traditional work culture on the English farm. Clearly informed by "New Woman principles of plain speaking and rigorous self-analysis,"[57] Hockin's *Two Girls on the Land* is an account of a gendered agricultural experiment that means to test and hopefully prove the physical "strength and endurance" of women on the land.[58]

As another second-generation member of the Arts and Crafts movement, Hockin too lives out and communicates the feminist, early green ideals in which she had been schooled. This liberatory project had been well underway before the war, informing both her work as a militant suffragette and her painting, which often features neo-pagan themes of dryads and nymphs, as in her "Pan! Pan! O Pan! Bring back thy reign again upon the earth!" (1914).[59] Certainly, the green thread that links her painting, neo-paganism, militant suffrage activism, and her land work is clearly the desire to intervene in a masculinist culture that resists equitable suffrage, land, and labor practices; food justice; and the fair treatment of animals.[60] Like Sarah Grand, Hockin is particularly offended that rural England is perceived as so much "scenery" for the amusement of the rich or the middle class.[61] And, like Grand, Hockin can also espouse eugenicist views that advocate for resources to be turned to the "fittest" of the human population. This emphasis surely informs her careful scrutiny of her own physical strength and her concern that she measure up physically and mentally to the grueling work she encounters. Her narrative of her time on a Dartmoor farm deploys the by now familiar early green and feminist discourses of the *aube de siècle,* pointing toward a disturbingly fit ecological modernity rising in a New Dawn.

Hockin's writing is both "performative and prophetic" in the tradition of New Woman narratives; she writes herself into being as a physically strong and competent woman with nearly each page.[62] Her example clearly seeks to prove to her readers that women can do manual labor and be leaders in the Back to the Land movement.[63] It is equally clear that she is committed to the more mystical or spiritual "Back to Nature" movement that led her to seek out intimate contact with and knowledge of "Nature" and that she seeks to communicate this to her reader as well as her visual audience.[64] In this way, Hockin's landswoman writing complements her now-much-neglected paintings. Her moorland, South Downs, and wooded, often moonlit landscapes show the vibrancy of plants, trees, grasses, and a minutely detailed knowledge of the complexity and diversity of woodland undergrowth and plant ecosystems. These landscapes are often animated by outlines of satyrs and dryads, transparent figures inhabiting the landscape. Paintings like "A Cobwebbed Woodland" (1912), for example, represent a dense, almost symphonic wood floor, alive with color—finely textured, variegated lichen and moss; leaves, fallen trees, and branches, all draped with cobwebs. Beneath the webs, Hockin hints at the vitality of the material world waiting to be discovered and revealed to those willing to look closely.

Initially, this aesthetic seems to have inspired her theosophist expectation that by "being with Nature, out under the open sky, with everything beautiful round and nothing to jar or disturb one's contentment, it would be possible to combine what one might call a state of religious worship, with work for one's daily bread."[65] Like the woman in "Imagination" and "Reality," however, Hockin finds that traditional agriculture, understood through "three seasons of hoeing," does not provoke the religious ecstasy or intimacy she was expecting; instead, it provokes the opposite—a "deadly sameness, its paralyzing effect upon the mind, and the intense physical fatigue which renders one insensible to any joy whatever in life except the relief of stopping."[66] To Meriel Talbot's distress, Hockin attributes this to traditional labor practices and argues that it is in the "the excessively long hours that lies the hitch. And if so, one cannot but feel, with Tolstoy, that if everyone did their share of 'bread-labor'—of producing absolute bed-rock necessities—that possibly the lot of those who are already overburdened might be lightened."[67] This was not the reading of traditional agriculture and the English farm Meriel Talbot was hoping to find, nor in fact is it what Hockin hoped to find in pursuing land work.

Although neo-paganism and the Back to Nature movement may sometimes seem ethereal, Hockin's memoir is grounded in sharp observation and experience of the material world and farm culture. Her artistic skill, political convictions, and beliefs combine in *Two Girls on the Land* to produce a detailed and intimate study of farm life and its politics, economics, and practices. Far from nostalgic, the book is dedicated "to the laborer on whose unceasing toil our ease depends," and this relationship, particularly between the city's reliance on the country population of unseen, low-paid farmers and laborers, is subjected to extended critique in the text. On the farm, Hockin, like Cockerell and Nussey, means to prove that women can contribute to the early green agenda that England should feed her own people. However, the methods Hockin uses to do so are as distant from the alternative market garden's cutting-edge practice of intensive culture as night from day. Seeking to prove that she can work as hard and well as a man on an English farm, Hockin comes to the conclusion that, overall, men and soil work too hard there. Labor practices are ruthlessly hierarchical, designed to exhaust the laborer to benefit the "Maester," the farmer.[68] Her constant evaluation of traditionally gendered and classed farm work is subjected to a steady analysis with the fresh eyes of one new to the work.

One of the problems with Hockin's new role, she finds, is its lack of innovation. She becomes schooled in traditional gender roles, to the point where her name and former identity fall away. She finds she prefers to call herself "Sammy" or "Samuel," rather than "Miss Something."[69] Living as Sammy and writing as Olive, she notes her need for a another partner, a woman friend she refers to as "Jimmy" or "James." But the pair find, ironically, that they have need for a "wife," calling upon Jimmy's sister, who comes to cook and clean for them. Sammy and Jimmy struggle to measure the limits of their women's bodies in men's roles against each challenge: exhaustion, influenza, weeds, headstrong horses and livestock, two broken fingers, and, ultimately, a desperate journey undertaken to move the farm's herd of sheep south during the unusually long, cold winter of 1917. At each moment of these trials, Hockin rigorously analyzes the role her body plays in her success or failure. Her conclusion that any inexperienced man might have found the work equally difficult freshly interrogates a primary New Woman objective: what Heillmann and Beetham describe as the New Woman's desire for "absolute bodily autonomy."[70] Through her own attempts at mastery of the tasks at hand, Hockin comes to the conclusion that no such absolute autonomy exists for anyone.

Hockin's task sets her up against a steep learning curve, which goes against the grain of her desire to "do any mortal thing I wished to do." Her realization of the complex mingling of human and nonhuman actants on the farm results in a humbling and deepened understanding of human limits (physical and cultural) and the power of matter itself: soil, weeds, animals, snow, ice, wind, water, and more. The result is less the triumphant narrative Meriel Talbot most "feverish[ly]" and "insistent[ly]" hoped for but found "never came" in the text, and more a deeper understanding of immanence.[71] What results is a horizontalist perspective of herself in the material world, a reduction in ego, and a critique and interrogation of the idea that any people, particularly the middle class, should be "free" of bread labor.[72] The struggle to reach these insights is viscerally communicated in the text through her detachment from her own body, her difficulty in accepting, for example, two broken "metacarpals" in service to the farm within only one year, and her vulnerability to pain and cold. While the memoir ends with illness and broken bones and her own admission of disappointment in herself, it blazes forth with new knowledge of contingencies, epiphanies into human physical strengths or limits, and dramatic, sudden understandings of human vulnerability, arrogance, and privilege rarely recognized but played out in the basic everyday act of eating itself. "OH, you—readers in furnished dining-rooms, who only know potatoes on your table baked and luscious in their wrinkled jackets," accuses Hockin, "little do you reck, as you help them from your dishes, of the toil and sweat and aching backs of those who labor over them the season through!"[73] Indeed, the impact of Hockin's honest analysis of Britain's food production and consumption can only be produced through her own journey to this understanding from a position of ignorance and confidence.

The memoir, then, is less the triumphant discussion of women's land work Talbot was expecting and more a record of the development of Hockin's ecological and social insights that are gained from studying where food comes from. What results is a call to dismantle the farm and replace it with cooperative work, fair hours, and a leveling of hierarchy.[74] Hockin's analysis proves that farm work could be dramatically lessened without the rigid gendered division of labor between "Maester" and "Missus," and without the prohibitions she sees against cooperative work among farmers. She notes that middle-class conventions delegate much of the work to the farmhand rather than the farmer or his wife. As soon as she arrives at the farm, she notices that the "Missus" never works outside the house and that others in the community look down on women who do farm work. Each task that Hockin takes on is subject to just such a class and gender analysis. Like other New Women writers, her own body becomes a "central battleground for contending ideologies" in a gender debate.[75] She seeks to test the limits of her own strength and, when it fails, she is relieved to see that men use the principles of "jujitsu," balance, or leverage, for example, to plough or to place heavy horse collars on the plough horses.

No physical challenge, however, is a more dramatic test in the memoir than the weather in the winter of 1917. At the hill-top farm's elevation on Dartmoor Hills, its valuable flock of sheep is particularly vulnerable to the freezing cold, which deprives the farm of running water. The only solution to an increasingly dangerous situation is to move the sheep on foot to a friend's farm further south, where the endless winter has already turned to spring. As the last healthy laborer left on the farm (second broken finger aside), Hockin volunteers to make the journey with the sheep, on horseback, accompanied by shepherd dog Shag. This epic tale, as Meriel Talbot writes, is a "vivid description of a dangerous and ghastly ride through a snowstorm, so dramatically told that at the end of it, one's own face was blue with cold, and stinging from the needle points of frozen snow which had been driving against it for hours."[76] Having undertaken the two-day journey to deliver the sheep to their temporary pasture, Hockin has no choice but to make the return journey with horse and dog. Climbing out of the warm valley where they have left the sheep in the care of another farmer, the team ascends the moors as a "blinding blizzard" gathers and the wind cuts "one's throat like a knife." The force of the elements attacks mare, dog, and rider until Hockin's "whole body seemed to dissolve with fatigue." Topsy, the horse, becomes team leader, making her own decision to "swing round." Hockin dismounts. No longer mistress of the group, she is "overwhelmed and beaten." The three "creep for shelter" into a hedgerow, "until the fury of the wind should be appeased."[77]

During this ordeal, Hockin thinks she is "about at the end of my tether. I felt that it was a physical impossibility."[78] Finally reaching the farm, in part because the horse recognizes the home road and moves forward, her return becomes an opportunity for Hockin to consider the full impact of her journey as a landworker: "With my weary mind and aching limbs I looked back at my former self, and wondered, amazed, as I remembered the young and enthusiastic applicant for work who had climbed that hill with such ease. I felt at least ten years older from the year's hard work—I dared not think of the day's work on the morrow. The same unending round would have to be gone through, but how to do it was a problem that seemed insoluble."[79] Comparing herself to a failing soldier, Hockin decides to plan her "defection." This proves unnecessary, however, as the farmer greets her with the news that the Army is sending them some soldiers to work the farm. "So that is how the end came. Our responsibility was over. Wife-Elsie in any case was leaving us; Jimmy was not allowed to come back, and I, with my broken finger, which was by this time becoming painful, was temporarily helpless. All our beloved animals—horses, cows, calves, and cats, we left to the care of the borrowed soldiers—and what happened after that we have never heard to this day."[80] This abrupt ending obviously displeased Talbot: "The result—the ultimate end of land work according to this book is this—these two girls break down in health, and their places are taken by two soldiers."[81] Bitterly disappointing, this conclusion defies Talbot's ideology on two counts.

On the one hand, Talbot insists, the landswoman, like the Angel in the House, should speak publically only of "the supreme joy of helping to make things grow—the love of the animals, the sense of freedom which the close contact with Nature brings you, the larger view of life, the realization of the wondrous beauty of the Lord—beauty of which should be, and are, a part, and which we have no more right to mar by ugly hearts or unhappy faces than the flower or the butterfly has the right to spoil its petals or its wings."[82] On the other hand, engaging women's bodies as an ideological battleground, Talbot brooks no admission of "physical failure" or illness as the result of land work. In fact, "the perfect happiness comes with perfect health" to be gained on the farm; the latter is "the secret of the mysterious fascination of work on the land" that should only energize, never enervate women.[83] Talbot has little appreciation for the insights with which Hockin walks away through her "ghastly" encounter with the storm. Hockin's unpacking of her own idealization of a Romantic topos of Nature in contrast to her new understanding of the materiality of the thing itself is not the preferred reading to which Talbot subscribes.

For Hockin, however, the experience on the farm provides crucial new insights, not the least of which is into the aggrandizement of the human ego. To most,

> man—oneself—is the center of the universe, and all else is " scenery!" All else revolves as subserviently and unmeaningly around him as did the sun and the planets in the days of Copernicus. There is no true happiness to be found in Nature, no real comradeship, until one can put away that all-important ego, and feel oneself of no more—and perhaps of no less—consequence than just any little toad-stool under a tree. What is man but a mite,

on a mite of a planet, revolving around a sun that is after all but one of the lesser stars of the myriads that make up the Milky Way! How, then, can he be of any more (or less) importance than the midge that dances for a day and a night over the stagnant pool? Why should he stride through the universe regarding all that is bigger, and all that is smaller than he—just as so much scenery?[84]

What Hockin has learned on the farm and in the woods is that humans are not necessarily at the center of life on earth. The erosion of the claim that she can do "any mortal thing I wished to do" is an ecological insight that acknowledges the agency of other things as well as sharp problems in human agricultural practices. While her realization is not a triumphant feminist victory on her terms, it is nonetheless the product of feminist inquiry, a situated knowledge that she arrives at through her own construction of an early green feminism.

At the end of her memoir, Hockin notes that "we may have been failures, but we did our best. We failed, indeed, at the end; but the task was almost superhuman. It would have taxed the strength and endurance of any man not brought up to the work from childhood, as the hardened laborers are."[85] Hockin notes that she cannot think of one middle-class man "who would have stood one week of the work that we did," and she also characterizes farming itself as an essentially exploitative, "decaying" business. The war should change such labor practices, not "merely . . . crush the enemy."[86] Hockin's conclusion is not happy, but enlightened and fervent: "To know so much of the laborer's life has been a rare privilege and one we shall never regret. We learned to know and to respect and to admire him; and we learnt how much the weight of the world rests upon the patient shoulders of the manual workers, the class formerly of all others to be despised, but now in the future to receive its due—in honor at least, if not in worldly riches. For ourselves, we saw the seasons through in all their changes of work and weather."[87] Hockin's struggle highlights the sheer physical challenges of traditional farming in England and sets out an agenda for reform of farm labor itself, a fuller participation by all people in the cultivation of their own dinner. Setting out to prove herself as a woman on the land, she runs into the thing itself in all its vibrant materiality, often vastly more powerful than she is. Taking shelter in a hedgerow, breaking a bone when a horse merely tosses its head of its own accord, subject to viral attacks of influenza, Hockin's body brings her to a horizontalist analysis that trades the New Woman's ideal of absolute bodily autonomy for a New Woman's ecology.

"On the Land": Rose Macaulay's Landswoman Poems

Hockin's reflections on the brutality of farming are pushed further into the realm of the postpastoral by Rose Macaulay in her striking suite of poems "On the Land." These document her experience as a landswoman and, in particular, as a noncombatant. Macaulay identified as a feminist and as a pacifist for much of the early twentieth century, working eventually "in the War Office, with responsibilities for exemptions from service and conscientious objectors."[88] Although her land work shocked some of her older relatives, much as Cockerell's had shocked Octavia Hill (an uncle writes of "Rose M now spreading manure!!"), she found the work "congenial. . . . Being outside in the open air gave her—as always—a sense of emotional freedom."[89] But "even as she enjoyed the egalitarian camaraderie and appreciated the freedom of working on the land, she continued to explore the ambivalence of her own feelings as a noncombatant."[90] After the war, she became an early woman member of the Peace Pledge Union, an affiliation she ended in 1940.

Although it has been said that "no one would confuse service in the mud of Norfolk with service in the mud of Picardy," Macaulay shockingly compares her service in the fields to that of men in the trenches.[91] Through this comparison, she develops a kinship between women agricultural workers and male soldiers on several levels. Like them she questions the official wisdom guiding her tasks, locating in these the absurdity of militaristic strategies that fail to take material obstacles more fully into account. In this way she enters the reader into a new consciousness that could best be described as postpastoral. Like Hockin, Macaulay experiences the leveling of human agency in her sometimes excruciating land work on inhospitable, long-fallow ground. She recognizes that agriculture is both a creative and destructive process in which people are deeply implicated and that the exploitation of plants and animals by humans descends from human tendencies to exploit other people.[92] On the other hand, Macaulay witnesses the agency of the material world itself both within and without the tilled fields she occupies. Sitting in a ditch alongside such a field, for example, she is overwhelmed by its biodiversity, perceived through her and her fellow landworkers' cigarette smoke during their break in "Lunch Hour." Macaulay's critique of monoculture involves a rethinking of its environmental costs and is a startling challenge to the sacralizing of the plough in English culture.[93] Her poems counter the Georgian pastoral poems associated with early Great War literature. Her raw, visceral writing on her land service demands that the reader see the bodies, elements, and agencies represented in them as immanent materialities with which humans are intimately entangled, with every bite of food they take. Even from its dedication to friend and editor Naomi Smith, Macaulay's *Three Days* recasts the "lovely and comic earth" as itself a writer awaiting the interpretive skills of an insightful reader. This earth is animated and legible, waving "wide hands" at the perceptive reader. It is a "witty," "strange" poet to be read by the right eye.[94]

However, in *Three Days* that rich material world has suffered. In dark, post-pastoral poems like "The Farmer's Boy," Macaulay makes the historical connection between war and agriculture quite clear. She envisions existing farm land as "naked country, without trees," a "tilled country, without dreams." "Bald and bare," she explains, "they do say there were forests here once on a day; But the great wars stole them away" to build ships.[95] She characterizes farmed England as conquered or vanquished; the landscape once commandeered to feed the navy timber is now a production site that feeds the war. Even in this dark vision of the English landscape, however, Macaulay's *Three Days* maintains a consistent emphasis on "hear[ing]" the earth, which speaks "a singing like to harps in my ear" as "like a ship at sea the wind goes" of its own accord and will.[96] In Macaulay's dedication, she gives the material world to her editor to read, asking that it be reconsidered, rethought for its own "rich jests." The latter claim is held in acute tension in the poems to come, which recall and interrogate purported divisions between nature and culture, the mind and the body, legible earth and text, combatants and noncombatants, growers and consumers of food. This tension is explored throughout *Three Days* and particularly in the collection's "On the Land" suite, which link the human practice of cultivation to power as it is dispersed between people, animals, the elements, and places. Ordered, almost madly, by superiors, to plant wheat on chalk land for which it is fundamentally unsuited, Macaulay's speaker knows that what she asks from the land is absurd. Whether or not she knows its full ecological cost, she senses the violence and damage wreaked by the plough in local places such as the chalk land of One Tree Field in Sussex, where she is set to work. Ordered to spread manure, neatly, "like marmalade on bread," she follows orders, knowing her work is futile. The manure, she writes, "must be all spread out ere the earth will yield / As it should (And it won't, even then)."[97] In this poem, the landwoman's knowledge is bracketed and subordinated to "orders" that nonetheless are effectively made to seem absurd.

Reading Macaulay's landswoman poems indicts the sentimentality surrounding women's "heroic" wartime food production. Planting becomes instead an often foolish or even obscene act that materializes embodied relationships to power. "On the Land" outlines the understanding that the agricultural countryside is not a place apart from culture, but deeply implicated in it. Macaulay's striking dismantling of the material earth from a powerful national myth begins with her unconventional analysis of conventional tasks for which the poems are named. "On the Land, 1916" includes five poems: "Driving Sheep," "Burning Twitch," "Hoeing the Wheat," "Spreading Manure," and "Lunch Hour," all of which develop a sophisticated commentary on her labor and presage other poems in the collection, like "Farmer's Boy" and "Dust to Dust." At the same time, most ironically, the poems represent startling insights into the richness of the local ecologies she recognizes for the first time. Like the "witty earth" of the dedicatory, they often represent the otherwise unseen vibrancy of the land itself in unexpected places—in ditches, for example, rather than ploughed fields.

The first four of the "On the Land" poems all begin with a gerund phrase that emphasizes the invasive nature of the landswoman's labor: "Driving Sheep," "Burning Twitch," "Hoeing the Wheat," and "Spreading Manure." Each poem expresses the knowledge Macaulay learns from each of these very particular agricultural tasks, set within harsh and exquisitely specific, lapidarian ecological settings. This new knowledge is gained in the face of certain paradoxes: her own inexperience contrasts dramatically with the immensity of her task, her privilege and education contrast with her incompetence on the land, her national task itself contrasts with the subjection of local places and local inhabitants, her vulnerability is both everything and nothing in contrast to that of the latter or further, the unimaginable vulnerability of other peoples and places of the world at war. Constantly attempting to represent her place on these discursive and material planes, Macaulay challenges the most treasured conventions of nature writing.

In the opening poem of the "On the Land" suite, "Driving Sheep," for example, she strikes a discordant note that challenges the pastoral tradition outright. The latter may be characterized as typically represented by a kind of "truancy," a retreat from the world of war and strife or of epic quests. The pastoral speaker abides in arcadian territory, enjoying "lowly, quiet life," as Sir Calidore does, for example, in "Venus's earthly retreat" in book 6 of *The Faerie Queene*.[98] If the "point of the pastoral life" is not to "escape from worldliness and its discontents" but to reflect upon those experiences, eloquently and in courtly language, with "self-awareness and nuance," from within the safety of arcadia, Macaulay represents arcadia as violated, even by her intrusive presence.[99] The poem begins with the women awakening in the dawn to lead the sheep from Chalk Field Pen to Grantchester: "Dazed with sleep, and numb." The sheep are meek, "unreasoning, blind" and each "takes its thought from the sheep next ahead" as the women parallel them: "Counting the sheep . . . we sway . . . into sleep . . . and trail along ... foolish as they."[100] In this context, with their strange new drivers, the herding seems mutual: the sheep and the women march "wild and pale, like the dead ... their sharp, scuttling feet" go, "strange, strange as dreams before day."[101] In this unnatural and nightmarish moment, however, the women, walking mindlessly with the sheep, also witness the sun rise in the countryside. They see "wide tides of gold surge, quiet and cold; The green rest turns deep blue."[102] Unexpectedly, the speaker is given a gift of intimacy with the material world: "There seems no sound in the world all round / But of horn feet and quavering cries / In the young, cold hour ... Like flame, like a flower, / The sun springs, huge with surprise."[103] Consistently, the harshness of the women's plodding and hard labor ("keep steady, move slow, we've three miles to go") is juxtaposed with the speaker's sudden and keen connection to the material world. It bursts in upon her, "huge with surprise" and thrilling, warming, and connecting to her something "old as the world," although at the same time, strange, eerily new as the war.[104] In moments like this, during the panicked march from the Chalk Field to Grantchester, Macaulay's poetry connects viscerally to the immanence of the material world and her own immanence in it as well. For the first time, she is both a "sheep" and a "comrade," a sister laborer, witness to the sunrise. The poem draws a disconcerting parallel between human and animal panic, a keen observation of a traditional agricultural task and sudden awareness of how that task is subsumed into the violence of total war. While the sheep are headed to the Chalk Hill Pen, either for shearing or for slaughter, the landswoman's own former "numbness" to her own immanence, to the knowledge of where warm wool and mutton originate, or what the sun looks like rising over a group of women and animals, both instrumentalized, is momentarily realized in a humbling epiphany of her own state of dependence upon the sheep, and their dual vulnerability as noncombatants ironically contributing to the war effort.

Macaulay's disorienting insight into the vitality of the material earth and her horizontal position as "sheep-woman," herding sheep and following orders, happens simultaneously, fracturing the soft glow of the pastoral with the development of an understanding of a dizzying rearrangement of vertical and horizontal relationships between culture and nature. Her questioning of her own role on the land continues in the striking poem to follow, "Burning Twitch." Here Macaulay explains how the brutal task of destroying the weed called "twitch" in an agricultural field reverses assumptions of human superiority to plants and educates landswomen in the resilience of the weed. "Couch" or "twitch" grass (*Elymus repens*) is a rhizomatic grass native to Europe and "anything but a favorite for the farmer for it has a slender, creeping rhizome ...

These long, creeping subterranean stems increase with great rapidity, and the smallest piece left in the ground will vegetate and quickly extend itself, so that it is almost impossible to extirpate it when once established in the soil, while its exhaustive powers render it very injurious to the crops. Its very name, Couch, is supposed to be derived from the Anglo-Saxon, *civice* (vivacious), on account of its tenacity of life."[105] In the formerly agricultural land Macaulay works, long turned over to pasture rather than to food growing, the invasive weed with its creeping underground stems has taken hold. Burning twitch now stands between the nation and starvation.

The crucial task, she explains, is to burn the twitch all the way through, for both grass and roots contain the invasive rhizomes which easily reestablish in a field. "[W]e kindle the twitch as the wind wills" Macaulay writes, "And burn it as we can and must. It smoulders, and crackles, and darkens, / And sulks between gust and gust."[106] The grass has dried in the sun and creates "blue smoke . . . drifting / Straight into our crying eyes," but the women must drive on forking "flame from bonfire to bonfire, before the windy blazing dies."[107] Still the twitch resists: "Twitch is like souls in hell torment, For it burns, but it never burns out. There is always some twitch left over, Though we, like the devil's rout, / run with our forks round the charred heaps / To prod them and turn them about."[108] By the end of the poem the women reach their human limit: "Our dim eyes are blind with crying, Our feet wrenched by baked earth-clods. . . . The twitch is gathered in from the plough lands, / And burnt up on midsummer day. But hell fire blown by wings of angels/Would not burn all the twitch away."[109] At the end of this poem, the twitch has won. Not only does the twitch seem empowered, but so do the elements of air and fire, which leave her gasping and blinded.

Stumbling, crying, and "wrenched," the speaker fleshes out the inadequacies of the term "on the land," which now seem a vast understatement. As these poems transform the meaning of the suite's title, they lend a disconcerting new perspective on the very boundaries the phrase assumes exist between the body and the land. Being "on the land" not only "wrenches" her ankles, but wrenches her consciousness from human arrogance to humility and grants her a grudging respect for a mere "weed": an agentic entity of the more-than-human world, strong, fierce, more resilient than humanity. Such encounters with the land produce humility rather than elation.[110] Neither her education, presumed quick intelligence, nor class privilege is a match for the rhizomatic agency of the twitch grass, which defies complete eradication and articulates itself within her very joints. The horizontal stretch of the rhizomes defeats any attempt at transcendence.

As Macaulay writes, she opens the surface of monoculture, and the national diet, up to critique. Following "Burning Twitch," "Hoeing the Wheat" is a shocking critique of bread and meat as the diet of victory. In the official discourse, both men and women were "eating for victory," and meat and bread were sacred foods upon which victory depended. While women and children at home should "eat less meat" and bread, the latter was crucial fuel for soldiers abroad.[111] Bread, meat's partner, is deeply symbolic even now in representing culture's transformation and refinement of nature. It is the mythical food of civilization, human ingenuity, and human dignity. But in "Hoeing the Wheat," the landswoman speaker critiques the merciless cost of bread to both the land and the farmer. As the women hoe the wheat field by hand from morning until night, the speaker identifies a mutual violence in both the cultural practices of ploughing and making war.

While burning twitch takes place in the heat of midsummer, hoeing winter wheat takes place "in wet wheat and cold." Accomplishing only twenty rows by noon, the clumsy and inexperienced recruits in the poem destroy not just weeds, "thistles and dandelions, And all the plaintains brood," but sometimes by mischance they hoe up something sacred: "A swathe of People's Food."[112] The long day in the field forces the narrator to reflect the far less labor-intensive eating habits of other animals. These "other beasts eat gladsomely," the speaker notes; "the earth's a table spread for them." "Man alone... in his insatiate greed, / Must break and cut the earth to bits / Before he can feed." With all else in the world to eat, "You'd think man would be satisfied— But oh, he is not."[113] To get bread, the women must fight weeds. By nightfall, "we've murdered docks and dandelions/Down fifty scorched rows." She wishes to be something else entirely, more than human, wiser and gentler. "I wish I were a rabbit, hid in a green lettuce bed ... Or any kind of animal who eats, but does not hoe."[114] The poem's conventional rhyme makes its startling insights into the violence of food production even more discordant and jarring. "Hoeing the Wheat" resonates with a full understanding of the cost of monoculture to the land and the women; it speaks more widely to darker forms of "man's insatiate greed," the breaking and cutting of the earth and bodies to bits in trenches elsewhere, a violence that seems incarnate and palpable, in the laborious, costly process of growing bread for "man" now at war.[115]

These issues surface and converge even more darkly in "Spreading Manure," which draws a now-explicit parallel between the Landswomen in agricultural trenches and soldiers fighting in trenches abroad. Ordered, unwisely, to grow crops on chalk meant to support grassland, the women in the poem know, quite like soldiers in trenches, that their service will not "yield" results.[116] The image of the women spreading manure in a rational, organized manner "fifty steaming heaps ... lying in five rows of ten" is an image of futility, dispensed by a rational, "cultivated" wartime authority that dictates what the women "must" do despite the land's very reasonable resistance.[117] The women's knowledge of the absurdity of their orders is ironically bracketed; they know despite their task to make "the earth ... yield / As it should ... (and it won't, even then)."[118] The women's parenthetical knowledge is irrelevant to the invisible authority that "drive[s]" them in all of the poem's imperatives: "The stuff must lie even, ten feet on each side, / Not in patches, but level—so." The fiercely laborious task, driving, spreading, "jerk it with a shoulder throw," is maddeningly precise for all its stupidity: "When the heap is thrown, you must go all round / And flatten it out with the spade." The next step is bizarrely domesticated: "It must lie quite close and trim, till the ground / Is like bread spread with marmalade."[119] Raw and darkly comic, the poem's central placement in the suite of poems highlights the speaker's ironic awareness, contrasting with the preceding poem's claim that "earth's a table spread for" life with "all the fruits of the woodland trees ... and eggs of birds, and honey of bees ... nuts ..." and even "little beasts a-running round / all ready for the pot."[120] In contrast, monoculture seems a "murder[ous]" undertaking based upon a theory of what the land "should" do rather than knowledge of what it *will* do "even then" of its own accord.[121]

With her images of jabbing, stabbing, and jerking, Macaulay's land work seems to mirror her masculine counterparts' service abroad. The work numbs her mind and even makes her mad. Her physical suffering is acute as "the north-east wind cuts and stabs our breath; the soaked clay numbs our feet. We are palsied, like people gripped by death, / In the beating of the frozen sleet."[122] Although she is not a soldier "sitting in the Flanders mud," she feels herself to be on a battlefield—not against people, but against a powerful force, the local ecosystem. In contrast, the trenches seem civilized; a dugout and a grave are at least "a six foot bed ... off the open land: A deep trench I could just endure." But "things being other," Macaulay not only must wrestle the elements; she "must stand / Frozen, and spread wet manure."[123]

Macaulay's image of futility and immobility while spreading wet manure is provocative on multiple fronts. The poem's title not only invites the shocked gender and class reaction the common reader, like her uncle, displayed—"Rose M spreading manure!"—but it also insists that this dark, postpastoral vision of food cultivation—"Spreading Manure"—be represented, become the subject of poetry. The poem extends the shock of the encounter that Hockin describes and the *Landswoman* represents. They too invoke the right to discuss an indelicate topic such as manure and explain the link between animal waste and human food production. But Macaulay extends her poem beyond the gendered and classed cultural borders of the garden text and into the sacred ground of the literary pastoral. In her poems, the green and pleasant land of the propaganda posters is represented as infertile, unyielding, and especially resistant to the plough and ill suited to the sacred crop of victory, wheat. The land, in turn, possesses its own kind of will, rhizomatic and vibrant. The poet claims in poems like "The Farmer Boy" that the land has long been commandeered or neglected for official uses. Indeed, there is no more affront to the sacralized national narrative of the Land Army than the truth Macaulay tells of the state of the land itself. In the poem "Spreading Manure," the titular act references not just the agricultural task, but the wartime spreading of propaganda about the viability and wholesomeness of the wartime landworker's task.

Like Hockin's situated knowledge, the wisdom that Macaulay develops on the land is particular to her wartime service. Sharing these keen insights with other women seals and shapes the camaraderie of the landswomen in the "On the Land" poems while it also, as in "Driving Sheep," expresses concern about their role in this official project. In the last entry in the collection, "Lunch Hour," Macaulay describes her comrades as "withdrawn for a little space from the confusion / Of pulled potatoes littered on broken earth." They rest in a "shadowed ditch, a peaceful circle / Of food, drink, smoke, and mirth." In contrast to the "broken earth" "littered" with potatoes, the ditch is vibrant and alive. Its smell is "hot and sweet, and heavy / With poppy flowers, and tangled with nettle-weed."[124] This ditch ecology contrasts sharply and ironically with that of the sparsely cultivated fields. The grassy ditch is inhabited by a cricket, whose chirping seems an "eternal question, / Like a thin tune on a reed."[125] The women view the fields they have worked as "littered trenches," which they now see as less vibrant than the ditch, which is alive both with human and more-than-human nature. Macaulay describes how "Blue tobacco-smoke drifted and curled about us; Its eddying wove for us a mystic screen. The field and its littered trenches dropped, and shimmered / In the clear gulf between / Real and dream."[126] Resting from labor, the speaker is enthralled with the imaginative divide or gulf between work and rest, between the women's and the ditch's vitality and the field's broken state, between their illicit smoke and the purported wholesomeness of the English field. The "gulf" that divides these two states of consciousness is a liminal space that holds an alternate beauty like the vibrant ditch, a beauty "strange, and thin, and far."[127] While the ditch and its occupants materialize in texture, sight, sound, and smell, the world outside their smoke screen seems "quiet and flat, as pictures woven / On old tapestries are." The ditch seems real and vital, the field unreal. The speaker addresses her companions directly as "we lay and laughed in the breathless noon-tide": "Your laughter, and your faces, burnt with the sun, Were as far and as near as heaven, and as mystic . . . / And the lunch hour was done."[128] The speaker's fleeting epiphany recognizes not just the ephemeral quality of the landworkers' intimacy, but references as well how much more there is to know of the land and of each other as women, an interrogative voiced in the sound of the cricket's "question."

This insight develops from the speaker's schooling in intimacy, in the formation of a "peaceful circle," protected, screened by a ritual blue cigarette smoke. It occurs apart from the carnage of their task, their labor in the "broken field." At the poem's end, the women rise "stiffly" and return to the "sunbaked trenches, And fl[i]ng the lifted potatoes into pails." As they work their vantage point shifts again: "And the earth stood out once more in relief and shadow, / Wholesome, like fairy-tales." The final lines of the poem return the earth to its three-dimensional proportion. But the purported "wholesome[ness]" of the fields now seems a fiction, a quality that is itself unreal: "like fairy-tales."[129]

Such moments resist Meriel Talbot's demand that agricultural work reify the Land Army's conservative discourse of wholesome gendered labor and the New Woman's vocational discourse to take all labor for her province. Instead, Macaulay's "On the Land" poems expose the violence of the landswoman's agricultural task and the dizzying new knowledge gained in service, including the knowledge of intimacy between women. In "Lunch Break," during which the women both eat the fruit of their labor and reflect on the violence of its production and wartime purpose, Macaulay celebrates the women's vision of their field from a ditch, through the smoke of their cigarettes, as their small harvest "litters" the earth like trash. In reporting their vision, Macaulay demonstrates the questions that can be raised by disturbing the most basic assumptions of pastoral; in a seemingly sustainable activity like farming she sees what Patricia Yaeger theorizes as a "dirty ecology," in which the means of a purportedly wholesome, life-sustaining system of cultivation is itself destructive. As the women rest, they exhale smoke and stub out their cigarettes. The image draws a parallel between their cigarettes and the produce "littering" the field, the soil of which they have "broken." At a much larger, almost inconceivable level, the work they do as growers is no reprieve from fighting; their food fuels violence on several levels, in both types of fields. Still, the vibrant life of the ditch is salvaged. In a dirty ecology people make "do with what they can salvage from" their human practices as the ultimate "waste-making" force on earth.[130] From the ditch, the women see that, in such an ecology, they are themselves "geologic agents . . . the main cause of change for earth" around them as they damage the soil, plants, people, in order to eat and, unthinkably, feed the war itself.[131] Yaeger argues for the value of such insights: "If we must scale up our imagination of the human, the consciousness of our scope and reach as species being, before we can hope to redeem the planet . . . this means owning up to the imbroglios we have made."[132] This "guilty recognition that we have the power to shatter our own universe is exactly the tragic recognition—a true anagnorisis—that we need to embrace . . . as species-being."[133] It seems that Macaulay reaches toward this recognition as she shares her view from the ditch, through the smoke she exhales. She perceives that "we practice a dirty ecology" through many agricultural practices in modernity, depleting soil to grow food regularly, and sometimes growing food to feed armies bent on world destruction.[134] It is fitting that Macaulay's speaker arrives at this insight during a break, a mythical "Lunch Hour," from which we can adjust our perspective on how we eat. In suggesting that the sustainability of the broken field is a "fairy tale," Macaulay addresses her own complicity with myth. Such representations, Yaeger urges us to see, are crucial to "reshaping a planetary epistemology," an ecological future.[135] Macaulay's poem suggests that this project may be begun through the situated knowledge of noncombatants, who see into the role food plays in world war.

A rare photograph of Macaulay at work with her companions captures just such an intimate glance into "a peaceful circle" of women formed around a broken field. Her cousin, Dorothea Conybeare, photographed her "and the other land girls. They stand in skirts and hats, holding hoes and pitchforks, grinning as Rose swigs from a bottle."[136] Three other women stand on the raw, broken field while a young girl sits at their feet next to a small, tangled pile of potatoes. The women, bundled in sweaters and hats against a winter skyline of barren trees, smile and laugh, leaning heavily on their hoes, while Macaulay, in a thick coat, drinks thirstily. The image, like other Great War images of women at work, speaks to the sheer exhilaration of women's groundbreaking work and camaraderie. However, the smallness of the group, the smallness of the pile of potatoes at their feet, the rawness of the field on which they stand, and, ultimately, the vastness in consequence and in desperation of their task also represent the insanity and chaos of both war and England's long history of food importation. Antithetical to propaganda images, Macaulay's work in the sacred national field is transformed to "the confusion of pulled potatoes littered on a broken earth," while a ditch and the women's camaraderie, their circle, nonetheless are both complicit and alternative spaces of vitality.

Reading Macaulay's "On the Land" poems as well as the letters, propaganda, photographs, and ephemera published in the *Landswoman*, it becomes clear that many recruits questioned their complicity with the war, critiqued the authority under which they labored, and coped with the vicissitudes of their work as growers who literally fed the war effort. They link human and ecological vulnerability in a provocative kinship, developing an acute awareness of the ecological absence of boundaries on earth between nations, between plants and people, and of the distinct historical and contemporary relationship obscured in modernity between growing, eating, and making war. Ultimately, landswomen's concern for their marking of the land, and in turn its marking of them, embodies new imperatives in environmental thought, encouraging readers to explore the agency of the environment, our interchanges with it, and the possibility for the formation of ecological knowledge within crisis. In documenting their encounters with mud, twitch, and manure, women landworkers came to know the vibrancy of the material world and the foolishness of those in authority who failed to account for it.

The conclusion of the Women's Land Army does not dignify that authority further. At the end of the Great War, women continued on the land until the following August, in 1919. At that time Meriel Talbot encouraged them to continue and "enroll for another term" as "your work is still needed."[137] In August 1919, a new editor of the *Landswoman*, Bertha Rayne, reported that a bill was before Parliament to offer land resettlements and cottages in England for all members of the Women's Land Army and armed forces, women as well as men. That bill did not pass, although a 1916 Settlement Act benefited returning servicemen. Members of the Land Army who had served at least six months were instead offered scholarships in a Free Passage program in order to enable their emigration.[138] In November 1919, Rayne announced the demobilization of the Land Army and expressed her satisfaction that women have now earned "a permanent place" in agriculture.[139] The *Landswoman* would continue until 1922. However, the nation would swiftly return to importation at the same rate as before the war, reinstating the economic conditions of the long agricultural depression which would persist until the outbreak of the Second World War. Institutions of higher education like Swanley and Studley Colleges continued on but were hampered by a lack of government grants, as was the endeavor of women in agriculture more generally.[140]

In 1939, the Women's Land Army was redeployed at the outbreak of war much more quickly than it had been during the first crisis. However, now the discursive image of the New Woman was carefully avoided. In language and visual images in national propaganda, women on the land were represented as children rather than women. The name of the new Second World War–era journal of the Land Army, for example was changed from the *Landswoman* to the *Land Girl*.[141] The visual and literary discourse of the latter no longer idealized mobile, educated, professional women with images of an elegant and angular New Woman in slightly masculine attire. Instead, the "Land Girls" were objectified as pinup girls, usually represented in soft, appealing sweaters and curled, bleached hair. When land work was fully mechanized by 1939, it became once again profitable and of interest to men, who returned to the work, restarting industrialized monoculture in England. Despite this turn, it is important to read the marks Landswomen made in resistance to the process of co-optation and redefinition of their work by the nation and to realize that some women remained on the land without national support and even in defiance of official regulation of their work. After the war, for example, in the absence of a national funding scheme, the Land Army helped fund a "women's smallholding colony" in Lingfield. It survived from 1920 to 1939 and is only now being studied.[142] In reading the work of women herbalists who, in many ways, were just beginning during the Great War, moreover, it becomes clear that the Land Army not only contributed to the preservation of England's biodiversity, but modeled a gendered rural alterity and ecological futurity that ecocritics have not yet begun to remember, record, and analyze.

Reading such work is especially helpful and timely now, as ecocritics think through issues around local food, globalization, and importation so long deeply entrenched in the most basic material practice, eating. Currently, England imports at least fifty percent of its food, while the production to supply ratio has declined steadily since 1988.[143] England stands on the cusp of a "tipping point," a disruption in the linear march of importation and food production caused by climate change.[144] Local farmers are still not given sufficient incentives to practice soil conservation and Brexit threatens to accelerate soil depletion if it becomes more expensive to import food. As soil becomes more vulnerable to runoff and erosion, which has increased due to climate change, not enough has been done to secure "high quality soil and . . . mak[e] crop yields more resilient to any potential shocks."[145] While just recently the Women's Land Army has been honored with a national monument, now is also the time to consider their analysis of food cultivation and consumption, ecological and social justice in food production. Their insight into how eating and growing food may exploit people and soil is valuable. Indeed, their wartime writing makes sense not just of Great War ecologies, but of our own.

5

WORKING RELATIONSHIPS

Ecological Futurity and the Herbal Revival

As the Women's Land Army was disbanding after the Great War, a number of women with a specialized skill set were forming a cohort that would thrive well into the twentieth century. These women, Maud Grieve, Eleanour Sinclair Rohde, Agnes Arber, and Hilda Leyel, became the leaders of the modern herbal revival in England. Together, they transformed public perception of local herbs from "almost inert" weeds to potent partners in both domestic and commercial gardens.[1] Developing a variety of new professions for women, this circle fostered the idea that English gardens could be both "*useful* as well as pleasing to the eye" in the modern world.[2] Together, they formed an herbal ecology that valorized both the importance of local biodiversity and women's newfound competence to participate in health care through "a higher form of gardening."[3] The latter was associated with intellectual as well as physical labor. Most important, and most controversial, in reclaiming and expanding scientific, cultural, historical, and practical knowledge of herbs in their gardens and books, the women of the herbal revival challenged the gendered distance between nature and culture, botany and medicine, that had widened since the Enlightenment. Their bold interweaving of plants and people imagined instead the intersection of biodiversity and modernity, plants and women's professionalism, to the benefit of all.

The resonance of this cohort's work extends from their own texts, plants, and gardens into popular culture, where it is explored through the herbal poison plots of Agatha Christie (1890–1976), herself trained to dispense medicinal herbs as an apothecary's assistant during the Great War. Through their educational programs, such as Grieve's The Whins Medicinal Herb School and Farm, women led the way in increasing national plant literacy by sharing both seeds and information widely with the public, encouraging "every household in the kingdom" to grow herbs for private use or "profit."[4] This rediscovery of herbs often verges on a kind of "euphoria," which is represented both in

Christie's work and in the texts of the revival.[5] There the potency, chemistry, rich history and folklore, fresh scent, subtle beauty, utility, and resilience of herbs are a source of "keen delight."[6] For the leaders of the revival, herbs raised an almost limitless sense of opportunity, particularly for educated women. Women of the herbal revival collected, shared, and bred plants, wrote and collected books, planted gardens, and opened businesses. As they did so, they critiqued extant environmental narratives of degeneration and extinction as well as dominant modern epistemologies that claim the superiority of industrialized products like patent medicines over plants.

Indeed, during the Great War women learned that local plants like "monkshood, chamomile, deadly nightshade, thorn-apple, henbane, purple foxglove, fennel, opium poppy and valerian" were "valuable medicine."[7] Such plants, moreover, were often alive and well in twentieth-century England. Despite the incursions of industrialization, they were "far from hovering on the edge of extinction."[8] It was only *knowledge* of herbs and plant literacy that had vanished. In herborizing expeditions, Maud Grieve and her students exercised their newfound skills in medicinal plant identification and mobility to locate plants in unlikely places. They found that herbs were resilient, still populating urban areas and thriving in the sandy soil alongside railroad tracks and fields, in woods, hedgerows, and ditches, and along the boundaries of long-abandoned gardens. The abundance of plants rediscovered in the early twentieth century takes a starring role in the writing of the herbal revival and informs the structure of a work like Grieve's *A Modern Herbal* or Lady Northcote's *The Book of Herbs*. These books reject the focus on particular spaces or types of plants common to Victorian gardening books and instead are structured like Old English and early modern herbals. Such older texts are catalogs that pre-date natural classification—the division of plants into categories according to Linnaean taxonomy. They list and illustrate plants alphabetically, as potent individuals, identifying them by their common names and detailing what they can do rather than what they are. The potentially infinite structure of the traditional herbal catalog appealed to women historians of science like Agnes Arber, who wrote the first scholarly study of the early modern genre in 1912, and to practical herbalists like Maud Grieve, as well as to aesthetes like Lady Rosalind Northcote and mystery writers like Agatha Christie. For each, the genre opened a door into a vibrant world of possibilities: medical, intellectual, aesthetic, and even criminal.

Ultimately, however, the willingness of women herbalists to share plant knowledge "without discrimination" became controversial.[9] Not only does Grieve's *A Modern Herbal*, for example, revive plant knowledge, but the text is performative in the tradition of New Woman writing. An example of a modern "female first," it establishes women's capacity not just to grow herbs, but to shape knowledge and culture itself. The book writes women into being as cultural authorities and shows how culture may form around the "vibrant matter" of plants, rather than the other way around.[10] Whereas once "botanical books ignored the medicinal properties of plants and the medical books contained no plant lore," Grieve includes both.[11] An entry on foxglove explains how the plant matters to the garden, the woods, to other living things like bees and animals, as well as how its alkaloids may be used to make digitalin, the heart medication. In *A Modern Herbal*, Grieve explains how to extract and dispense these alkaloids, including cited material from the trade journal of her wartime allies, *Chemist & Druggist*.

Reviews of the book note the latter with alarm. Some complain that she shares "professedly medical" information with the public indiscriminately.[12] In explaining how to cultivate, handle, and dispense potent plants safely, Grieve crosses a gendered line laid down between the garden and the pharmacy by physicians, pharmacists, and chemists. Indeed, even as Grieve's work was championed during the Great War by the distinguished chemist and University of London professor E. M. Holmes, the medical profession more generally was working to increase the distance between garden and medicine both rhetorically and materially, framing herbs as inferior and seeking to synthesize plant alkaloids so as to distance the pharmacy from plants. Anxieties about an incursion into medical authority culminated in the "sinister" 1941 Pharmacy and Medicines Bill, which would strictly curtail herbalism in England and do great violence to plants and people.[13] Supporters of the bill denigrated medicinal plants, dismissing them as "dandelions and cloves" or "ordinary things," in a rhetorical move that sought to ally plants with the realm of nature and magic rather than modern pharmaceuticals.[14] At issue was access to authority over the *British Pharmacopoeia*, an official text that represented the masculinist realm of modern science. The result of this struggle was to harm both women and biodiversity.

The rich concentration of potent plants and women poisoners in Agatha Christie's interwar work, then, certainly engages anxieties over women's newfound authority over plants, which became controversial in the 1930s. Between the wars, Christie explores this intersection of plants and women, modernizing the earlier figure of the "New Woman Criminal" to do so.[15] The intersection of the latter with the kind of utopian euphoria the women of the herbal revival create is a potent combination in Christie's work. In her story "The Herb of Death" and her novel *Five Little Pigs*, she places herbs in the hands of irresponsible, "very modern" women, allying them with a volatile nature.[16] The conjunction of women and plants in the latter texts suggests that as the figure of the New Woman ages between the wars, it remains an often destructive force, distilled into a "blazing" anarchic vitality even near mid-century.[17]

By that time, professional women herbalists, gardeners, and scholars far exceeded Ruskin's vision of the Angel in the Garden. Books like *A Modern Herbal* suggested that women could gather, grow, and process herbs themselves to control and promote their reproductive and general health. Literary and horticultural collaborations between women absent male authority from their work as women write each other into being as experts, "leading specialists," a "circle of illustrious" but now often forgotten women.[18] The latter provide an apt final subject for this book, as they trace the intersection of early green and New Woman literature and culture well into the twentieth century. The discourse and narrative strategies of the latter helped men like E. M. Holmes and women such as Grieve or her editor Hilda Leyel to advocate for herbalism as a women's profession, particularly in the face of war, which was to increase the need for medicine and women's independence in tandem. Indeed, as professional herbalists, women like the entrepreneurial Leyel, who edited Grieve's landmark *A Modern Herbal* and owned the herbal emporium Culpeper House, greened the sexual anarchy of the New Woman in disturbing *new* ways, raising questions about the intersections of women's autonomy, plant potency, consumerism, and the masculinist field of medicine.

This bid for an ecological modernity begins as a form of women's aestheticism, with Lady Rosalind Northcote's *The Book of Herbs* (1900). Her admiration for the cultural and aesthetic importance of old herbs in contemporary England is developed in the scholarly studies of the early modern and Old English genre of the herbal by Agnes Arber and Eleanour Sinclair Rohde. They claim, respectively, that the archaic genre of the herbal matters both to the history of science and to human history. The genre is one of the very few historical records of "what plants have meant to people" and vice versa.[19] Maud Grieve advances their work by modernizing the herbal genre, retaining its traditional collage-like structure, which includes the cultural meanings and "modern scientific uses" of herbs in *A Modern Herbal*.[20] Grieve's book, a veritable "speaking garden" or cornucopia of plant and medical knowledge, put herbs on the map again in England. Although Grieve raised concerns about making medical knowledge available to both the "serious student and common reader," her book is still in print, even digitized, and still the definitive modern herbal it set out to be.[21] Indeed, many of the plants now ubiquitous—the hybrid Seal lavender, for example—may be traced back to Grieve and her students. Their work is still visible in plants, gardens, and books, which, even now, foster biodiversity and plant literacy in England.

Indefinite Boundaries

When he learned of Oscar Wilde's conviction for gross indecency, publisher John Lane of the Bodley Head was in New York City. From there, he cabled London frequently, urging his partner to pull Wilde's books from their list and fire Aubrey Beardsley, the designer of *The Yellow Book*, who was strongly associated with Wilde. While in New York, "oddly enough," Lane ran into Lady Rosalind Northcote at a house party.[22] It is perhaps there that the two began discussing her publishing in a new series on practical gardening. Clearly seeking to rebrand the Bodley Head after Wilde's conviction, the press soon came out with its first book in the series, *The Book of Asparagus* (1901), followed by Lady Northcote's *Book of Herbs* in 1903. While the former text was met with scorn by Max Beerbohm, who demanded to know if once "flowery Vigo Street" was to become "the kitchen garden of the Muses," a shrine "of carrot and cauliflower?" Northcote's book was well received.[23] *The Spectator* declared it stood "on a different plane from its fellows. . . . It has much more in it than 'practical gardening.' It has distinct literary value." The reviewer notes that "the subject, of course, lends itself to a literary treatment" and "there is a great store of curious learning in the book. . . . The series is most distinctly enriched by this latest addition to it."[24] Certainly not a practical guide, *The Book of Herbs* follows in the earlier tradition of the Bodley Head; it is an aesthetic text that genders its "curious learning" and claims it for women.

By the end of the nineteenth century, the very obscurity of herbs, their link to the occult and magic rather than to reason and science, made them ripe for reclamation by female aesthetes like Lady Northcote. Una Ashworth Taylor's 1896 short story "The Seed of the Sun," for example, characterizes the reputation of these plants at the fin de siècle. In the tale, a young flower seller is given a seed, "brown and shriveled." She is told the plant will grow "strange leaves," and she may see these but not the blossom. She seeks out an aged physician who knows "many things and has many books; in his garden many strange and rare and health-giving herbs flourished." But even he cannot identify the seed. As the plant grows, it saps the life of the young flower-seller. "Beware," the doctor warns her: "There are herbs which are for health-giving—and there are also herbs which breathe out strange maladies." The latter seems to be the case in the tale, and the girl dies soon after the plant blooms. Taylor's "The Seed of the Sun" associates herbs with antiquity, obscurity, and toxicity. The anxious search to identify the herb in the tale reflects and constructs the cloud hanging over herbs after a century that celebrated the wares of the story's protagonist: colorful, often-imported flowers. Taylor's tale reveals the limits of plant literacy after a century of importation, but it also raises the stakes of plant identification, making a bid for the importance of the latter as well as its exquisite pains and pleasures. Both the archaic material qualities of the seed and herbal expertise reveal why herbs were intriguing subjects for women aesthetes and scholars who sought to redefine the terms of knowledge itself.[25]

Seeking to know the unknown, the hidden gem, the obscure pleasures and rigors of an arcane discipline such as herbalism, women aesthetes like Lady Rosalind Northcote combined a wider interest in more-than-human nature with an interest in much-neglected herbs.[26] Northcote's early twentieth-century reenvisioning of the early modern genre of the herbal, *The Book of Herbs* (1900) begins by asking much the same question as "The Seed of the Sun": "What is a herb?"[27] But her answer is a critique of the question and a rejection of Linnaean taxonomy itself. Although Northcote has heard "many definitions" of the term, none have "satisfied the questioner." It may be "fairly safe to say," she writes, "generally that a herb is a plant, green, and aromatic and fit to eat, but it is impossible to deny that there are several undoubted herbs that are not aromatic, a few more grey than green, and one or two unpalatable, if not unwholesome." Northcote attenuates that line of inquiry, refusing to accept the botanical epistemology of taxonomy as a path to knowledge or insight, and rejects the formation of knowledge through the botanical practice of "natural classification." The classification of plants according to relationships based upon their descent from a common ancestor, she says, is limiting: "So no more space shall be devoted to discussing their 'nature,' but I will endeavor to present individual ones to the reader," who may then form their "own idea of a herb" from "their collective properties" or powers. In the tradition of the aesthete, Northcote is clearly bored with the limits of accepted knowledge, which studies what things are rather than what they do. She is interested in the potency and properties of plants. Each plant she discusses is described in the tradition of the medieval herbal text: according to its unchanging physical properties (color, fragrance, leaf size and texture, etc.) as well as its chemical properties (reactivity to heat or water, healing or antimicrobial activity, toxicity). Reclaiming a deliberate archaism in the tradition of fin de siècle aesthetes, Northcote deploys a model of plant identification that comes from the traditional Old English or early modern text in order to point out that "the boundaries of a herb-garden are indefinite."[28] Indeed, as a writer, Northcote benefits from the early modern conceit she deploys in her title, which itself erodes divisions between books and plants, exploring both as a type of literacy that expands the limits of the imagination. Deploying the historical alliance between books and botany established by early modern and classical authors, she too "sets the authority of poets" almost on a par with herbalists, using them as they did to "confirm matters related to their habitat or medicinal virtues."[29] The ecological story Northcote tells in *The Book of Herbs* deploys this deliberate archaism to reclaim all "we have forgotten" and "much that would be profitable to us" to know about plants in modernity.[30] As she references past and contemporary authorities on herbs in her mimesis of the collage-like structure of the early modern or Old English herbal, Northcote, like all the women studied here, includes women authors who provide her with a historical precedent and modern incarnation of women as authoritative practitioners and keepers of herbal knowledge.

Northcote's *Book of Herbs* writes women's authority into being, planting that quality deep in the aesthetic literary herb garden she places at the nexus of a wide variety of polymathic knowledge systems. Appearing somewhat ironically in the Bodley Head's "practical gardening" series, *The Book of Herbs* suggests that regaining knowledge of herbs might be more interesting than cultivating them. The archaic rhetorical form of the catalog is certainly more compelling than the standard structure of a Victorian garden text ordered around the design and placement of plants in the garden. The catalog structure allows the plants "virtues," complex histories, subtle, green or grey beauty as well as their fragrance, flavor, and medicinal potency, to stand out. Each chapter is organized not around garden spaces, but around the different uses of herbs in the present or the past, in medicine, in folklore and magic, in animal husbandry, and in design or cosmetics. Like the herbal, *The Book of Herbs* is richly illustrated, and it includes literary and historical references to herbs in a wide variety of texts from Shakespeare to Elizabeth Gaskell. One chapter alone, "On Growing," addresses cultivation.

Herbs in Northcote's text are living reminders of a lost woman's plant culture, thriving but unstudied, just outside the back door and still visible in early modern women's records of their household herbs or in Charlotte Yonge's poetry. Tantalizing ruins of herbalism's material culture remain extant in obsolete antique equipment like herbal stills, which she has photographed for inclusion. Assembling knowledge through objects, literary allusions, and folklore, Northcote inducts the reader into a sensuous materiality of plants, a pleasure gained through knowledge that begins with plant identification and the ability to distinguish herbs from mere weeds. She shares her passion for their intoxicating scents and transformative properties. Her revival of earlier women's herbalism resembles what Schaffer notes is a tendency of New Woman aesthetes to reconcile "domestic skill and connoisseurship" in women's aestheticism.[31] In fact, Northcote also means to reclaim gender authority in the herb garden entirely, to "transfer" herbs "from the patronage of the blue serge [a male gardener's uniform] to that of the white muslin apron [worn by women]." "Recovering" this tradition is an idea "we recommend" to "the consideration of our lady-gardeners," says Northcote.[32] Northcote means for the herb garden to be a woman's place, under female authority, and designed by her for her pleasure.

For Northcote, reviving herb gardening requires regaining esoteric knowledge and a new aesthetic appreciation of what had become unfamiliar, even strange, plants. The Victorian garden itself must be transformed as well, no longer a regimented display of gaudy color, but now a far subtler, sensuous experience of scents, textures, subtle colors, literary histories, and potencies. To illustrate the potential of the latter, Northcote includes striking original photographs of rough-textured herbs placed, like flowers, in vases, and reproductions of antique wood cuts and ballads (see fig. 11). She refers to the delights of scent—"The Neglected Sense," as explored by well-published connoisseur and collector of aesthetic Japanese objects, Edward Dillon, with whom she shares a fascination in a game the object of which is correctly naming the scent of incense or fragrant wood burning. Referencing Edward Dillon places Northcote, in her first book, in his company. However, like *The Yellow Book* and George Egerton's *Keynotes*, *The Book of Herbs* presents itself as a new kind of book, the chapters of which present a potent alternative culture, a "book of herbs" that marks the intersection of literature and the imagination with the materiality of plants. Such knowledge might set one apart from the crowd. While male aesthetes had claimed floral ecologies as their own, reshaping them so that they appropriated, in particular, singular, large flowers—sunflowers, lilies—as badges of their alterity, Northcote reclaims the smaller and subtler "magic" and "delights" of herbs.[33] Understanding not only their appearance but also the "properties" of these plants, their pungency and vitality as fresh culinary herbs and the powerful medicinal qualities released by their essential oils, and knowing how these impact health, energy, and vitality when consumed, makes the Victorian flower garden and its culture of bloom, despite its bright colors, seem inert, lifeless, and domesticated by contrast.

Indeed, Northcote's *Book of Herbs* brings alive the vibrant essence of plants, evident not only in the essential oils that give them their power, but in the lively role they once played in material and literary culture. Her discussion of distilling herbs and edible flowers transforms the value of plants from

Figure 11. "Rosemary," from Lady Rosalind Northcote's *The Book of Herbs* (1903). (Author's collection)

their mere "show" to their "substance," which partners with human life in a powerful alchemy through the cultural practice of distillation. What excites Northcote, like other herbalists, is the ability to transform a plant from solid matter into a pure liquid "substance," capturing its materiality in a new form, a process that often revealed a woman's skill and knowledge of plants and her authority in the realm of medicine. Elixirs and tonics might contain a season, suspend or control time, transport the plant's vitality to another time and place. Northcote turns to Shakespeare's Sonnet 5 to prove the pleasures of plant distillation, in which a plant's "substance still lives sweet," "a liquid prisoner pent in walls of glass" in winter.[34] In this way, the herb garden truly has no boundaries. Its beauty lies not just in the eye, but also in the plant's "effect." In this sense, the pleasures of the herb garden are not just spatial but transcorporeal; the garden enters the human body not only through the eyes or nose but through the lips and very skin as tonics and syrups, salves, ointments, and food. Northcote's use of the catalog structure moveover implies that the knowledge of the uses of herbs is itself beautiful, limitless, and ever material and healthful as well as intellectual and abstract. This knowledge defies the separation of nature and culture and is a great yet accessible pleasure that, once lost, can now be regained through study and practice nearly anywhere:

"In the expert knowledge of herbes what pleasure still newed with varieitie? What small expence? What security? And yet what an apt and ordinary meanes to conduct men to that most desired benefit of health?"³⁵ In the tradition Schaffer identifies of the female aesthete who valued slim, less-expensive eighteenth-century antiques or wildflowers over objects of weightier antiquity or hothouse flowers, the cultural abjection of herbs makes their recognition all the more special.³⁶ Furthermore, the knowledge of the historical role of herbs in human culture opens the door to new medical epistemologies. As Rohde would explain in *A Garden of Herbs,* herbs once maintained health proactively, and health itself was framed positively rather than "negatively" as the "absence of disease."³⁷ A preface to such work, Northcote's aesthetic *Book of Herbs* playfully but steadily lays the ground for the herbal revival, encouraging future women to claim realms of authority and competence.

Keen Delights: The Scholarship of Agnes Arber and Eleanour Sinclair Rohde

As an early collector of herbal texts herself, Northcote was instrumental in renewing interest in the manuscript and early print culture of Old English and early modern herbals during a time when such texts were dismissed as mere curiosities. Her work was furthered exponentially, however, by historian of science and Newnham College fellow Agnes Robertson Arber and by author/gardener Eleanour Sinclair Rohde. They sought to reclaim Old English and early modern herbal texts as crucial cultural knowledge, a "chapter in the history of botany" and a crucial record of the interdependence between plants and people.³⁸ Arber completed the first scholarly study of early modern printed herbals, publishing *Herbals: Their Origin and Evolution—A Chapter in the History of Botany 1470–1670* with Cambridge University Press in 1912. Rohde then went deeper into the past, publishing *The Old English Herbals,* a study of herbal manuscript culture, with the Medici Press in 1920. Rohde followed this text with *A Garden of Herbs* (1922), *The Old English Gardening Books* (1924), and others such as *The Scented Garden,* which shifted trends in garden design by introducing herbs into modern culture. Arber's landmark text on early print herbals elevates the importance of the genre of the herbal by placing it within a progressive scientific narrative, published under the imprimatur of a university press. At the same time, her book establishes her as a new kind of woman. In analyzing the herbal as intellectual history, Arber was among the first degreed women to write "books about books."³⁹ Her maiden name on the title page was followed by her married name and academic titles "Agnes Arber,

(Mrs. E. A. Newell Arber) D. Sc. F.L.S., Fellow of Newnham College, Cambridge and of University College, London," establishing both her academic status and her respectability now marshaled to uplift the study of common plants like herbs and women's capacity to shape knowledge, culture, and history.

Although the overarching academic thesis of Arber's text is her claim that herbals contribute to the more advanced science of botany, her "keen delight" in the epistemological richness of her subject is clear.[40] Like Northcote, Arber is among the first British people to argue that the "commonness" of herbs should not disqualify them from study.[41] As a scholar, Arber researches the conventions and generic features of the European and English herbal, its alphabetical listing of plants and their virtues, and the genre's use of coded illustrations, which often feature plant roots as well as insects whose sting or bite the herb may treat—an important visual aid in the promotion of plant literacy.[42] She is interested in the genealogy of the herbal, in collaboration and competitions between herbalists, and while she is intrigued by what herbals contribute to botany, she is critical of what botany failed to take from them. Arber's text notes the "older herbals" portray "the plant as a whole, including its roots . . . of special value from the druggist's point of view." In "modern botanical drawings," the recognition "of the paramount importance of the flower and fruit in classification has led to a comparative neglect of the organs of vegetation, especially those which exist underground."[43] Arber argues that botanical epistemologies have short-changed plants, causing their uses to fall out of the definition of plant knowledge itself. Moreover, in early modern herbals "plant drawing, as an art, may be said to have reached its culminating point . . . treating the plant broadly as a whole, and not laying more stress upon the reproductive than the vegetative organs."[44] Post-Enlightenment botanical illustrations, she finds, obscure medicinal information, and their use of Latin obscures access to plants for the common reader. Arber recognizes that the rise of botanical discourse, in celebration of the complexity of natural classification of plants, had made plant description impenetrable to those uninitiated in Linnaean taxonomy. In contrast, the herbal's description of a plant as "thick as a man's arm" not only was clear to the common reader, but brought both plants and people closer together, as comparable, even horizontal entities, both equally vibrant if of different flesh.[45]

The material, imaginative, and intellectual pleasures of early modern plant culture suggest an interconnection between people and plants that Arber finds beneficial to both, quite apart from the efficacy of the science of early herbalism. Although not "quite free from superstition . . . their enlightenment was quite remarkable."[46] Arber ends her book on a wistful note, pointing out that "as time went on, the *herbal*, with its characteristic mixture of medical and botanical lore, gave way before the exclusively medical *pharmacopeia* on the one hand, and the exclusively botanical *flora* on the other. As the use of homemade remedies declined, and the chemist's shop took the place of the housewife's herb-garden and still-room, the practical value of the herbal diminished almost to vanishing point."[47] It is just this gendered herbal "vanishing point" that women during and after the Great War seek to redress.

"Irrevocably influenced" by Agnes Arber's *Herbals*, Eleanour Sinclair Rohde made "her historical and practical interest in herbs and unusual plants "her profession throughout her life."[48] Among the first gardeners trained by Grieve, she was to write ten books and cultivate many herbs. Rohde attended Maud Grieve's The Whins Medicinal and Commercial Herb School at Chalfont St. Peter, Buckinghamshire, in 1916, and by May 1919, she was assisting Grieve in the creation of the first herb garden to be exhibited at the Chelsea Flower Show. Rohde seems to have been an early member of the British Guild of Herb Growers from about 1918.[49] She should be "credited with much influence in reviving the growing of neglected herbs, and recalling and passing on the tradition of monastic gardens in cultivating fragrant and culinary plants. She was a member of the committee advising Mrs. Hilda Leyel on the management of the House of Culpeper," and she designed one of the first major modern herb gardens at Lullingstone Castle, Kent, in 1946.[50] The latter was, like Vita Sackville-West's herb garden at Sissinghurst, also begun in the 1940s, an important mark of the success of the herbal revival. (It is all the more unfortunate, then, that Rohde's walled herb garden was replanted in 2015 to hold the imported "world collection" of the current landowner, exotic "plant hunter" Thomas Hart Dyke.) In her writing, Rohde's position is that no national history is complete without an understanding of local plant life and its intersection with human culture.

At the beginning of her book *The Old English Herbals*, for example, Rohde argues that the genre of the herbal should be considered as important as literary texts. "[I]n the whole range of Saxon literature," she writes, "there is remarkably little mention of plant life. The great world of nature, it is true, is ever present; the ocean is the background of the action in both *Beowulf* and *Cynewulf*, and the sound of the wind and the sea is in every line. One is conscious of vast trackless wastes of heath and moor, of impenetrable forests and terror-infested bogs; but of the details of plant life there is scarcely a word."[51] In herbals, however, "we find what plant life meant to our ancestors."[52] Moreover, while Old English literature recorded the deeds of great men, herbals recorded "what plant life meant to ... everyday folk. ... And even in these days to understand what plant life means to the true countryman is to get into very close touch with him."[53] Rohde argues that herbal texts become a way of regaining knowledge of people as well as places. She describes such knowledge as a national "birthright."[54] For Rohde, plant literacy is crucial to valuing both people and biodiversity; in fact, the interests of the two, nature and culture, are intertwined rather than opposed to each other.

Framing national history as natural history, Rohde seeks to regain a historical understanding of the potency of local plants on several levels. As well as powerfully medicinal, plants are resilient, capable of playing a complex role in English culture over hundreds of years. People have been cut off from their knowledge of ecological interdependence, which she now sees as destroyed by the monotony of "suburban life." The latter has "separated the great concentrated masses of our people from their birthright of meadows, fields and woods," she writes, and few have interacted with a "wilder" Nature: of "Nature, in her untamed splendor and mystery, most of them have never had so much as a momentary glimpse."[55] Rohde seeks to restore those glimpses through the study of old English herbals, and, as her books are performative, they encourage people to engage not just old herbal texts, but still-vibrant herbs themselves, through gardening and herborizing. Unlike Arber and Northcote, Rohde made her living as a gardener; she maintained a nursery in which she cultivated long-neglected herbs and preserved special specimens, designing humble and grand spaces such as the herb garden at Lullingstone Castle.

For Rohde, "old books" about plants should not be regarded as "quaint, if not ridiculous. This attitude seals a book as effectually and as permanently as it seals a sensitive human being. There is only one way of understanding these old writers, and that is to forget ourselves entirely and try to look at the world of nature as they did. . . . They believed . . . that natural forces and natural objects were endued with mysterious powers."[56] These old herbals and the still-existing plants they described were survivors. They contain not just plant information, but an approach to bodies that in Rohde's mind was pushed out along with the industrialization and professionalization of the medical trade. Rohde argues that the latter "usurp[ed] the place of" herbs in culture and pathologized the body: "We have come to look upon health as the mere absence of disease," she writes. "With us it is a negative thing; but the word 'health,' with its cognates 'holy,' 'whole,' 'wholesome,' has a positive sense, and the old herbalists were never wary of preaching the use of herbs, not only to cure, but also to keep one in perfect health."[57] Rohde makes remarkable connections to the effect of herbal culture not just on health, but on perceptions of health that were shifted with the recasting of herbs and plants as inert, less effective medicines in modernity. Not an occultist, Rohde nevertheless understands that animist perceptions of plants created a human connection to them. The model of medical care that existed in such societies, where medicine was environmentally accessible and interwoven with growing things, seems eminently just to her, giving people full understanding of both human bodies and plant bodies.

Rohde envisions plants and plant knowledge as a powerful alternative epistemology in which all had access to healing. Knowing that the same plants—digitalis, coltsfoot, horehound—exist then and now creates a model of an ecological modernity that is the antithesis to contemporary wartime apocalyptic and extinction narratives. Picking up where Arber left off, Rohde connects the plants of the past with a still-vital ecological present and future. Although old herbal knowledge, like the sharp scents and sometimes bitter taste of herbs, might "strike our modern minds as quaint, or even grotesque," the plants' colors, utility, humility, and helpfulness to humanity seem, after the war, preferable to the showy display of more "stately gardens." The "fashions in large gardens have changed," Rohde argues, and it is "assuredly" the herb-growing "cottage garden which is characteristically English."[58]

Writing through the dual ecological and human suffering of the Great War and the English agricultural depression, Rohde encourages her reader to turn to a garden more somber in color but rich in comfort. Among her five epigraphs to *A Garden of Herbs* is John Ruskin's admonition that "all the wide world of vegetation blooms and buds for you; the thorn and the thistle which the earth casts forth as evil to you are the kindliest servants; no dying petal nor drooping tendril is so feeble as to have no help for you."[59] She quotes Pater: "The worship of Demeter belongs to that older religion nearer to the Earth, which some have thought they could discern behind the more definitely national mythology of Homer." Pater's Demeter moreover "knows the magic power of certain plants cut from her bosom to bane or bless. . . . She is the goddess of the fertility of the earth in its wildness."[60] Waking this sleeping goddess after the unthinkable losses of the Great War is Rohde's task and teaching the power of plants is her skill. However, Rohde's neo-paganism reconstructs the fertility of the earth without linking women to the damning heteronormative floral discourse that is by now all too familiar in saccharine, disciplinary garden texts like *The Secret Garden*. At the same time, herbal power to "bane or bless" poses a bittersweet alternative to the floral ecologies rejected by antiecological modernist narratives. Her dedication of *Old English Herbals* to her brother, who was killed in the Great War, seems particularly appropriate to the materiality of the subject matter.

Rohde's work, and the work of many of the herbalists of her generation, counters the "defoliation" of much contemporary literature.[61] In "The Waste Land," for example, T. S. Eliot seems to find the return of plants unbearable. Flowers are obscenely sweet; lilacs and hyacinths dare to grow out of graves: "The roots that clutch ... What branches grow / Out of this stony rubbish?"[62] In the face of the flower bearing "Hyacinth Girl," the returning soldier cannot speak. He is blind, "neither / Living nor dead and I knew nothing."[63] But Rohde, a grieving sister, a war survivor herself, tasked to professionalize and financially support her home and parents, turns to growing things to do so. The things that console her are reasonably plain, often bitter, but potent, resilient, and both ancient and modern, familiar and strange. In her books, herb cultivation and processing is differentiated from the cultivation of flowers, vegetables, or corn: "Herbs ask so little and they give so much. All that the majority of our common herbs want is a fairly poor soil (the poorer the better for the aromatic herbs) and plenty of sunlight."[64] After invoking the garden lore of five esteemed men in her epigraphs to *A Garden of Herbs*, Rohde pointedly asks: "But what do we know of herb gardens?" Effectively identifying the knowledge gap into which her book enters, she declares "For we use so few herbs, and those we have relegated to an obscure corner of the kitchen garden. It is a little difficult even to imagine," she continues, "a time when 'vegetables' occupied only an insignificant part of the herb garden, and a still earlier time when both the flower garden and the vegetable garden were nonexistent, and the herb garden reigned supreme."[65] Among Rohde's insights is her dismantling of the idea of the garden as a stable entity unto itself. Instead, she shows gardens to be culturally constructed and historically contingent spaces, impacted by importation, fashion, and, not least, war. In envisioning the need for gardens to change in the aftermath of the latter, she reflects both the creativity and skill of her mentor, the herbal "luminary" Maud Grieve.[66]

Maud Grieve: "Uncultivated Ground"

At the center of the circle of the women of the herbal revival, Maud Grieve deploys key aspects of New Woman literature and culture to shape a gender-equitable, green future for England. Her massive, two-volume *A Modern Herbal* defines contemporary herbalism, revising the tradition of the Old English and early modern herbals to include cultural, historical, and cutting-edge medical and scientific knowledge. Tempering New Woman utopianism with that figure's emphasis on the rationality and intellectualism of women, Grieve argues that the logic of laissez-faire liberalism is no logic at all, for it ignores the bounty of English plants and English labor. Grieve's model of the educated woman herbalist challenges the nonecological and unsustainable trajectory of Victorian import ecologies, which she indicts as fundamentally illogical: there is *"no reason,"* she argues in one of her earliest pamphlets, that herbs might not be grown for profit in England.[67] Nor is there any reason why women should not be growers. Herbs used by "by every household in the kingdom," Grieve writes, thrive in England's temperate climate.[68] They might become "a considerable source of profit," as they "might be much better grown in this country."[69] Promoting the commercial and domestic planting of herbs in England assured that women would have work and herbal culture itself would be revived. That culture promoted ecological and social ideals such as plant literacy, local biodiversity, and access to employment and affordable healthcare, particularly for women. *A Modern Herbal* implicitly and expressly promotes the notion of women's bodily autonomy and competence as healthcare providers and professional herbalists in modernity.

Like others of her circle, Grieve moves the gardening advice text into the public realm, guiding growers and herbalists from the private, domestic garden to the farm or herbalist's shop. Once in the public domain, woman herbalists destabilize the authority of masculinist, industrialized medicine *and* challenge the "death of nature," which the latter augured.[70] Grieve resisted that narrative and its contribution to dominant narratives of degeneration and extinction. Even the long subtitle of her book (*The Medicinal, Culinary, Cosmetic and Economic Properties, Cultivation and Folklore of Herbs, Grasses, Fungi, Shrubs and Trees with All Their Modern Scientific Uses*), and the sheer weight of the seven-hundred-plus-page text communicates the exuberant vitality of plants and her refusal to equate defoliation with progress and modernity itself.

Like the rest of her cohort, Grieve challenges the gendered division of culture and nature that is at the heart of the masculinist industrialized medical pharmaceutical complex. Including both botanical and medicinal information, her book is a performative intervention. It writes biodiversity into being and, at the same time, writes herself and her cohort into being as experts, holders and keepers of cultural authority and vast scientific knowledge. As Grieve reveals her comprehensive knowledge of hundreds of plants, she explains their potency and the vitality of the people who plant, care for, and harvest them. This intersection between plants and people creates a vision of England as a vibrant productive garden fostering interdependence and intimacy between human and plant life, culture and nature. The book demonstrates how the interests of people need not be separated from the interests of an abstract and objectified Nature. At the same time, the relation between modern people and plants is neither simply destructive nor exploitative but far more complex.

The ecological story that runs through *A Modern Herbal*, then, refutes Victorian declensionist narratives and the contemporary narrative of "the death of nature" in modernity.[71] Each of the plants Grieve describes still exists, and she indicates that there are many more in need of study. Because such common plants had been of little interest to botanists, they have the impact of a new discovery in Grieve's book. In fact, as Kay Sanecki says, upon publication, *A Modern Herbal* exploded on the "calm waters of the horticultural and botanical world," where it was hailed as a masterpiece.[72] Grieve's massive catalog of living plants and their modern uses insists that even postwar contemporary England is vital, populated by countless plants then not even recognized *as* plants, but rather classified as weeds by the population. Moreover, all sorts of urbanized and industrialized land in England, the sandy soil around railroad tracks, "damp waste ground," ditches, and "rubbish heaps," Grieve argues, are not dead zones, but full of potent green life.[73] When Grieve began her work, "mint, parsley, sage, thyme and marjoram" were in use, but she notes that "many others . . . deserve more attention than is generally given them."[74] Grieve covers the history, cultivation, and use of nearly one thousand plants in her illustrated herbal. Her text unfolds, as John Fowles says, like a massive "paper garden," planted with seemingly new and exciting species, which had been there all along.[75] What results is a survival story in which herbs authorize women's mobility in culture and counter the idea that modernity is fatal to Nature and that Nature is fatal to women's full participation in culture. The text entirely refutes the then-still-entrenched idea that women were less capable of scientific and rational thought than men.[76]

Transforming herbs from "inert" to potent and vibrant plants with a purpose transforms the gendered discourse of gardening as well.[77] The return of herbs heralds a new kind of national garden tended and sustained in futurity by working women who benefit from doing so. Educated, professional women should do the "higher" work of herbalism, cultivating, processing, dispensing, and researching herbs.[78] These women, Grieve argued, were especially well poised to do the "thoughtful," responsible work of herb gardening and herbalism "by co-operative action."[79] Grieve opposes the interlocking elements of free trade, monoculture, importation, masculine economic authority, and leisured, female consumption in her performative text, demonstrating in her maturity her own success and her great knowledge coming into clear sight with each entry. Her book constructs a vision of herself as the example of an expert professional woman, offering others the advice they need to follow in a book that contains a potentially limitless modern green world. That utopian vision interlocks with the possibility of an edible, biodiverse English landscape that grows food for people and provides habitat, nectar, and fodder for beneficial insects and animals. Indeed, the possibilities of herbs imbues Grieve and her peers with a kind of euphoria in the possibilities of herbal ecologies during the herbal revival. Grieve's utopianism, however, is always balanced by her maturity, rationality, and deep practical experience. She writes as an experienced gardener.

Departing from taxonomic botanical epistemologies, like Northcote and Rohde, Grieve focuses on relationality and intimacy between plants and other living things, including, of course, people. *A Modern Herbal* charts the endless complexity of people's entanglement with plants. The catalog will never be complete, as new varieties of herbs are always discovered, evolving, or created. Her narrative of the astounding proliferation of herbs and the diversity of plant life in England alone disrupts the terrible end story of human dominance and ecological apocalypse of her and perhaps our time. Certainly, her work disrupts the ecological narrative explicit in our own concept of the Anthropocene, which often shapes a view of the passive earth impacted by agentic humans. While I do not by any means mean to suggest that human impact has not been egregious and devastating, recontouring and damaging the earth severely, I do mean to suggest that this particular vision of human agency tends to obliterate earth's capacity to act itself.

As Heise argues, environmental stories of species loss and extinction, the death of nature as measured in redlining of extinct species, from the beginning of the nineteenth century onward, have dominated ecological discourse in sometimes exclusive ways. This is particularly true in plant studies, where the study of the evolution of new species and resilience of old species has been most neglected.[80] Grieve's perspective on plants is particularly valuable, as it traces their response to industrialization, framing them as resilient, exuberant actants rather than passive beings bowed under the weight of human activities.[81] Plants in Grieve's work jump the garden gate, act and react, move, and spread, suggesting that "there are instead always a swarm of vitalities at play" within and between human and plant culture.[82] The fertility of the foxglove alone is "almost incredible," Grieve writes, "a single Foxglove plant providing from one to two million seeds to ensure its propagation."[83] Such a plant also engages in complex mutual interactions with other beings. It is "a favorite with bees . . . much visited by other smaller insects, who may be seen taking refuge from cold and wet in its drooping blossoms on chilly evenings, yet no animals will browse upon the plant, perhaps instinctively recognizing its poisonous character."[84] Grieve's interest in bees and the plants that support and benefit from them is constant throughout the text, explaining the mutualist role herbs play in sustaining themselves and pollinators. Her discussion of the pollination of the foxglove provides perhaps one of the clearest examples of ecological mutualism in existence.

Grieve's discussion of the "visits" of the bee to the plant, its use of "the lower lip of the corolla" as "an alighting platform," and the "almost incredible" number of seeds produced from this cross-species cooperation, imagines not just the fecundity of Nature, but the complex, erotic ecological interaction between actants in the environment.[85] Her mutualist narrative manages to be both "stately" and sensuous, countering dominant narratives of competition, predation, and parasitism, and establishing how bee and flower sustain terrestrial ecosystem function.[86] Grieve is not the first to note that a native plant like the foxglove is a "favorite" for native bees.[87] However, the information takes on new meaning when she moves on to discuss the interactions between the plant and people in the past and present, all of which work to regain the plant's potency and human interdependence with plants. The structure of the entry makes the narrative of these life-giving contingencies between species precede human interaction with the plant, suggesting that people too are involved in the complex ecological story of the foxglove. Her use of the native English name for the plant and her explanation of its historical role in English culture links ecological and human history while her explanation of the plant's activities and potency borders on a vibrancy that is almost agency.

Of course, Grieve does not claim that plants are sentient, but she does disrupt the anthropocentric trajectory of environmental impact itself, showing that the resilience of plants is agentic and, hence, not only do human beings not act alone upon the earth, but plants have more capacity to act than they are usually allotted. The ongoing fecundity of herbs despite industrialization then proves that while humans consider themselves the active subjects in the world, they are not alone in determining the future. Herbs, in their resilience, actually dissolve this vertical hierarchy, showing how other living things may be actants if not sentient agents, as they have the capacity to "animate, to act, to produce effects dramatic and subtle."[88] In this sense, even industrialization is not simply an action but "a trans-action, and any act is really but an initiative that gives birth to a cascade of legitimate and bastard progeny."[89] "Garden escapes" and "volunteers" embody the resilience and mobility of plants long after humans have stopped tending them or are even able to identify them.

To see this green world, Grieve is bent on regaining plant literacy of common English plants long overshadowed by the botanical study of exotic plants.

In an entry on the herb henbane, used for pain relief during the Great War, Grieve first teaches her readers to see the plant, recognizing it as a potent but prevalent entity, rather than as "a weed of cultivation."[90] Reversing botanical discourse, "commonness" is valorized. Henbane, she writes, is valuable *because* of its prevalence and despite its preference for bastard locations somewhere half way between culture and nature. It "has been found wild in sixty British counties, chiefly in waste, sandy places, by road-sides, on rubbish heaps and near old buildings, having probably first escaped from the old herb gardens. It is frequently found on chalky ground and particularly near the sea."[91] Not only does Grieve revive the local names of plants, often, as in the case of henbane, linked to their properties, rather than privileging Linnaean Latin names, but she marks all such plants as (POISON) in the text, alerting the reader to their power. Grounding knowledge and evidence of that power in no less a local cultural authority than Shakespeare, Grieve jogs cultural memory of the plant's potency with a reminder that it is likely the deadly "juice of cursed hebenon in a vial" used to poison Hamlet's father.

Like Northcote before her, Grieve makes powerful allies of Marlowe and Shakespeare, Culpeper and Gerard, in reclaiming the potency of herbs throughout *A Modern Herbal*. At the same time, these early modern authors are a foil and preface to her modern expertise and to the important new scientific information she shares on the effects of and proper medical use of plants. Speaking from the unimpeachable position of her wartime service, Grieve explains authoritatively that henbane is a "valuable remedy, either as an anodyne, a hypnotic or a sedative." It will "take effect when other drugs fail." The plant, Grieve explains is the active ingredient in "Twilight Sleep" for which it "has of late come into use . . . causing loss of recollection and insensibility. Hyoscine is also used to a considerable extent in asylum practice for the treatment of acute mania and delirium tremens." The tiny amounts of the plant to be used assert its power as an actant: "Powdered leaves, 2 to 10 grains. Fluid extract, 2 to 10 drops."[92] The vibrancy of a plant like henbane is regained by its power in both nature and culture, which are interwoven in the entry.

Making this information available to the reader puts healing powers out their back door and on their bookshelf and provides that information as the country recovered from the trauma of the Great War. It is clear that Grieve imagines plants playing a large role in that recovery. Her long and complex entry on "Lavenders," for example, shapes a new vision of England as a new kind of garden, growing the plant of a peaceable future. Grieve's entry "Lavenders" is unusually long and complex, one of the most substantial entries in *A Modern Herbal*. From the beginning, her entry is comparative and competitive; English lavender is superior to French, Italian, or Spanish lavenders. In Grieve's mind "nowhere [is lavender grown] to such perfection as in England."[93] Grown under cloud cover rather than the constant sun of warmer climates, it has a "softer" scent. It is "much more aromatic and has a far greater delicacy of odor than the French, and the oil fetches ten times the price."[94] As delicate as its scent is, so is the plant hardy: "English lavender is more robust in habit than the French plant" and "is of fairly easy cultivation in almost any friable, garden soil."[95] As with all of her discussions of herbs, Grieve argues that local plants made to be consumed by local people are superior to imports, which must travel many miles to reach England. Lavender is impacted not just by transport but by soil, time, and conditions of gathering, rapid or slow transport to the still, distillation method, and weather. Because of this, local plants have an advantage: "English Lavender is always entirely harvested in under a week, and the flowers are distilled on the spot."[96]

For Grieve, lavender was supplanted by the war and it may now return. Once cultivated near London, the expansion of the city pushed the plant out as far as Surrey. "During the war," however, "the cultivation of Lavender was still further diminished to give place to food crops, so that in 1920 not more than ten acres under Lavender cultivation could be stated to be found in the whole of Surrey, though some of the oil is still distilled in the neighborhood, and the finest products continue to be described as 'Mitcham Lavender Oil.'"[97] Grieve argues that this situation can be remedied in a country now at peace. Her seven-page, minutely detailed entry on lavender makes it clear that the plant is linked to quality of life, national economic and ecological recovery, and a peaceful land. As she depicts its return to England, she imagines it thriving in ideal conditions: "The plant flourishes best on a warm, well-drained loam with a slope to the south or the south-west. . . . Protection against summer gales by a copse on the southwest is also of considerable value."[98] The ideal lavender farm will attend to each aspect of the plant's life, for it "requires very careful consideration." Directions for cold harvest days, windy harvest days, sunny harvest days, and wet harvest days are all given with the goal for "making the most refined Lavender oil," which will be "very superior to foreign oil of Lavender."[99]

The result of this labor and skill is pure herbal pleasure and a jewel of a plant of great value. From June the plant's spikes will be covered with oil-bearing "esters disseminated throughout all the green parts of the plant. . . . From this time onwards . . . the esters commence to concentrate in the flowering spikes" and "can be distinctly seen by the naked eye in brilliant sunshine, the tiny oil globules shining like little diamonds. The delicacy is completed by the concentration of the esters during the following month."[100] Grieve's pursuit of excellence in the cultivation of English lavender and her ability to compel her reader's interest in each detail constructs the plant as a potentially perfect being and the gardener as its witness and beneficiary. At the same time, Grieve's meticulous instructions construct her ethos as a "luminary" herself, the epitome of the woman gardener shaped by her own robust labor and rationality.

In her vision, the New Woman is at work pursuing a "higher form of gardening," deploying her rational thought and physical autonomy. This scene displaces the landscape garden maintained by male gardeners as the ideal English garden that represents the nation. *A Modern Herbal* transforms the latter into a work space run by women for profit. Grieve's vision of the ideal English lavender was to materialize in her student Dorothy Hewer's work.[101] Inheriting Grieve's knowledge and her stock after the former's retirement, Hewer, with the help of her assistant Margaret Brownlow, cultivated the now ubiquitous "Seal lavender." This plant was named for their woman-run school and herb farm, Seal Farm. The latter fulfilled Grieve's vision of the power of plants to transform women's lives and the power of women's lives to transform plants in a mutually beneficial fashion. Envisioning a future in the company of plants, Grieve's vision of an ecological modernity centers around the assemblage of plants and people around a variety of challenges. These include the devastation of war, the social, epistemological, and ecological losses of importation, and the underemployment of women. Her work is timely indeed, suggesting that ecological solutions may arise from the situated knowledge of particular historical crises.

"Professedly Medical"

Grieve's book was greeted with acclaim and outright joy from many reviewers, who recognized its rewriting of the genre of the garden book. Many saw that it established women's capacity to shape knowledge itself, intervening in epistemological divides between culture and nature, medicine and botany. However, some reviews of the book note the latter with alarm. In sharing medical information with both "the serious student and the common reader," Grieve crossed a gendered line laid down between the garden and the pharmacy by botanists, physicians, pharmacists, and chemists.[102] While the latter were welcoming herbs grown by women during the Great War and into the Second World War, they were also working to increase the distance between garden and medicine, both rhetorically and chemically. By 1941 and the introduction of the Pharmacy and Medicines Bill, Parliamentary testimony from the pharmaceutical industry would classify herbalism as mere superstition and plant-based drugs as "panaceas" and "nostrums," throwbacks to pre-Enlightenment England.[103] Advocates of the 1941 bill claimed that herbal treatments were ineffective "quack remedies," which may also be dangerous.[104] As "women herbalists held their ground as leading specialists" at the time, they were likely to be most heavily impacted by new regulations, which were accompanied with a push to link herbalists to the abject realms of reproduction, magic, nature, and ignorance.[105] In Parliamentary debate, contemptuous language struck at the very heart of the herbal revival, dismissing the value of "ordinary" plants such as "dandelions . . . cloves . . . things like that," while herbalists considered the same to be "among the most valuable of herbal medicines."[106]

Grieve's editor, herbalist and author Hilda Leyel, was among the chief defenders of herbalism from this bill, which was enacted in 1941. She wrote her book *The Truth about Herbs* (1943) to counter the damage done by the act and to resist the stereotypes and aspersions it promoted against herbalists. In her herbal emporium, Culpeper House, she had long worked against stereotypes that allied herbalist businesses with "small shabby room[s] with dirty windows . . . all suggestive of mysterious and probably evil practices."[107] Her shop was light and bright, designed to maximize the potential of the New Woman's consumer culture and the figure's association with cosmetics and health. Like forerunners Northcote, Rohde, and Arber, Leyel had worked to preserve and curate antique herbal manuscripts and books as a source of medical and social history. In 1927 she had helped form the Society of Herbalists, and her business had been thriving for years by the time the 1941 act became law. Her Baker Street shop was fashionable, while at the society Leyel also made her library of antique herbal texts and manuscripts available to members. It was here that she also began to offer herbal treatments to her clients.

In her response to the act in her book *The Truth about Herbs*, she claimed that her business was "competing with the most up-to-date methods and theories of modern scientists . . . within a few hundred yards of Harley Street and Wimpole Street, the Citadel of Orthodox Medicine." Arguing that her work was popular with "highly sophisticated modern Londoners," she defied the Parliamentary discourse that identified her products as "nostrums."[108] Her patients had "no superstitious beliefs in herbs at all," nor had they been lured in by false advertising.[109] Her practice had succeeded through word of mouth while her shop benefited from its celebration of local plants for local people. She had striven to make Culpeper House "extremely English in character," using "the old familiar names of many English flowers and plants" and selling "elixirs, distillations, and lotions . . . of our ancestors" as well as customized herbal treatments.[110] Just as her practice as an herbalist as well as a purveyor of cosmetics had begun, however, the 1941 Pharmacy and Medicines Act banned herbalists from making medicinal elixirs, ointments, distillations, decoctions and more from fresh, whole plants. Herbalism was effectively criminalized and it was not to revive again until 1968, when the act was repealed. This was an extreme step not taken by other countries such as Germany and France, which regulated herbalists, and in England, it meant human *and* ecological harm. The act injured the present and future cultivation of herbs in England, for without a need for fresh, whole plants, many could again simply be imported dry. It also curtailed the perception of herbs' potency, the possibility of human/plant interdependence, and plant literacy, as it reduced needs to mainly culinary herbs. Just on the verge of returning to growing fresh local herbs commercially, England found the demand decrease. A return to importation of cheaper products from France and Germany was inevitable, impacting not only people but English biodiversity as well.

While the popularity of herb gardening was to remain deeply entrenched in England after 1941, the injury of the 1941 act to the herb farming and herbalist industry was considerable. The banning of herbalism and the sequestering of the use of whole plants to licensed pharmacists effectively pushed popular knowledge of local plants, a key factor in the preservation of biodiversity, out of the hands of the common consumer or reader. Poised to regain an understanding of the interdependence of people and plants, the British public found a barrier erected between themselves and plants despite the textual knowledge they had so recently regained through Grieve's *A Modern Herbal*. Even deep knowledge of the latter owned by women like Grieve and her editor Leyel might now be suspect, while practice from its pages might be criminal.

This gendered struggle over plants is marked in the work of Agatha Christie. Reading the herbal revival back into Christie's texts raises the ecological stakes of her female poisoners and their intersection with the phenomenon of the New Woman Criminal. As her texts engage the historical debate over who should be trusted with plants, they gender the risks of ecological interdependence between plants and people in modernity as well as "modern" women's authority over the latter.[111] The herbal revival as well provides still-definitive contemporary tools with which to contextualize Christie's representations of plants more widely, beyond their mere associations with poison. Grieve's *A Modern Herbal,* for example, allows for a fuller understanding of contemporary and historical meanings of herbs in Christie's texts, contesting the limits she ascribes to plants and resisting her plots' process of defoliation, their undermining of the potency of "leaves and berries."[112] At times Grieve's knowledge reveals the biases in Christie's reframing of plant knowledge. As the most widely read texts in the English language, Christie's mysteries not only reflect, but shape perceptions of plant knowledge even now. They too vie to define "The Truth about Herbs," merging anxieties over human interdependence with plants into anxieties about women's capacity to think rationally, mirroring and refracting the contemporary debate that empowered the medical profession, and, finally, moving women and plants over to a lesser state of nature at great cost to both.

This representation of a struggle over women's ecological knowledge reveals a powerful trend at the approach of the Second World War to erase New Women's achievements. The latter, however, do not go down without a fight. In fact, the "sheer blazing vitality" that accompanies Christie's representation of women and plants is redolent of contemporary debates, which make the New Woman a compelling figure in Christie's interwar mysteries.[113] Hence, the glamorous, resourceful New Woman Criminal makes a late appearance, long after she first appears in late Victorian crime serials and early crime film. Christie's texts complete a trajectory begun earlier in English fiction when "cosmetics, lurid posters, vivid advertising . . . represented 'newness' for a culture obsessed with modernity, and they employed the female criminal to embody and explain the shock of modern life."[114] Her crime fiction of the era of the herbal revival references an increasing need to contain the New Woman's sexual anarchy through detection and government regulation in the future. Even estimable detectives like Miss Marple and Poirot encounter difficulties when "modern" women have access to potent plants like foxglove, belladonna, or hemlock. Regulation then suddenly seems salutary and safe by contrast.[115]

"Putting Men First" in "The Herb of Death"

Published near the release of *A Modern Herbal* in 1931 and the passage of the British Pharmacy and Medicines Act in 1941, Christie's texts "The Herb of Death" (1932) and *Five Little Pigs* (1942) engage historical anxieties over the intersection of women and plants between the wars. The structure of the mystery "The Herb of Death," for example, exposes, step-by-step, "who" and what people and plants "really are."[116] The text clearly seeks to reclaim plant knowledge in the tradition of the herbal revival. When this knowledge is regained, however, it is reframed in opposition to the gendered advances in plant literacy and science made by the latter. In fact, in the tale, women, in particular, even the tale's gardener narrator, are shown to be misinformed about the potency of "leaves and berries" in general.[117] The process of detection reveals that their assumptions about plants are based in myth and that their plant knowledge is insufficient to solve the crime at hand. In the text, women's knowledge, represented through the gardener Mrs. Bantry's assumptions, is overturned by male expertise. The latter takes the shape of the medical man, Dr. Lloyd, who dismisses the general idea that any plant's "vital principle" exists in its raw state: "These ideas of eating poisonous leaves and berries are very much exaggerated," he states.[118] The real power of plants, Dr. Lloyd explains in a disquisition on foxglove leaves, is known only to a "very few people," chemists and scientists, capable of the skill and knowledge through which they may extract the plant's alkaloids.[119] Dismissing the vitality of leaves and berries as folklore, Lloyd's new scientific information has the effect of undercutting the potency of plants and women's knowledge and authority in the tale and garden.

In "The Herb of Death," Dr. Lloyd's explanation seems perfectly reasonable; however, historically and ecologically, it is problematic on several counts, which makes Christie's tale a troubling example of a false divide between gardening and science. First, Lloyd discounts the importance of leaves themselves to the quality of a medicine like digitalin, the alkaloid of foxglove that works to regulate the heartbeat. His recommendation of the title "The Herb of Death" is ironic, for the tale proves it is *not* the fresh plant, but a chemist's alkaloid drug version of the herb that is the murder weapon. The leaves of the plant are a red herring and something of a cliché, a joke, like his other alternate titles: "the poisoned bulb, the deadly daffodils . . . !"[120] As far as the doctor is concerned, potency lies in the laboratory, not the garden.

This claim, however, is incomplete. The strength of the alkaloids extracted from foxglove leaves is dependent upon the gardener's rational cultivation and gathering. Maud Grieve explains that "no leaves are to be used for medicinal purposes that are not taken from the two-year old plants, picked when the bloom spike has run up and about two thirds of the flowers are expanded, because at this time, before the ripening of the seeds, the leaves are in the most active state."[121] Grieve's entry on the foxglove concludes with a two-page quotation from the journal *Chemist & Druggist* that explores the nuances of the varying strength of the plant's leaves and shares formulas for medicinal use of the plant, explaining how to extract the alkaloid, a relatively simple process that Lloyd implies is inordinately complex, known only to "very few people."[122] Not only is the quality of the drug digitalin dependent upon the age and condition of the leaves, but the knowledge Lloyd assumes necessary to make the drug is less exclusive than he imagines or than Christie represents. Just one year before "The Herb of Death" was published, Grieve had shared detailed instructions for the use of the plant with the common reader and gardener.

Gardening in Christie's tale, however, distances the materiality of plants from science, and women from knowledge and rational thought itself. Mrs. Bantry has great difficulty organizing her thoughts well enough to play her part in the Tuesday Night Club. For example, she is told not to speak from her head, but from "where your heart lies," in her garden.[123] From there, she must be heavily prompted to cross over into the realm of culture through storytelling. Other women in the tale are consistently characterized as lacking reason and insight. Even the name of the murder victim, Sylvia Keene, associates her with both nature and excessive emotion. She is assumed to have been poisoned by consuming fresh foxglove leaves she is thought to have picked herself. These were mistakenly stuffed in a duck by an equally "stupid" female cook, who received the leaves—thought to be sage—from Keene.[124] Dr. Lloyd rules out this cause of death through his confident explanation about the lack of potency of fresh foxglove leaves.

Even the estimable Miss Marple may solve this crime only through the new pharmaceutical knowledge contributed by Dr. Lloyd. Sir Ambrose, it seems, has poisoned Keene with his heart medicine. The story then registers unwitting female participation in a man's plot with a man-made drug. Moreover, women's alliance with plants—their admiring of flowers, picking herbs, cooking them, growing gardens—allies them with the material space of nature and places them intellectually in the dark. Just as Mrs. Bantry herself has difficulty constructing a narrative from her head as opposed to her "heart," so do other women fail to "think" rationally or possess deep knowledge or skill in the use of plants, as Dr. Lloyd and the murderer do. Sir Ambrose's letter of confession comes as a surprise to Mrs. Bantry. Indeed, Dr. Lloyd and Sir Ambrose work together to instruct women in the story: the latter explains his terrible deed, the former explains the science of it. Sylvia never suspected herself as "she was not perhaps, very clever. In fact, rather stupid."[125] Throughout the tale, men's intellects and their cleverness in extracting vital principles from plants are truly potent crime solvers and heroes. "I don't even like women," Mrs. Bantry says. "I like men and flowers."[126] Mr. Bantry thanks his wife for "for putting men first."[127] His comment might apply to the narrative trajectory of Christie's tale as well. Both put men first ideologically at the cost of the vitality of plants and women. While plant knowledge is regained in the tale, as it was historically between the wars by women like Maud Grieve and arguably Christie herself as a war-time apothecary, "The Herb of Death" undermines the potency of plants and allies them and women with a less agentic nature rather than a vibrant contemporary herbal culture. Ecological interdependence between people and plants more generally is undermined as well. Christie's mystery contributes to the defoliation of literature and culture in modernity and, at the same time, undermines women's knowledge of plants in direct contradiction to the historical rise in the same. "The Herb of Death" then reasserts the distance between culture and nature and masculinist visions of a triumph over nature by science, if not the death of Nature itself.

As in *5:20 from Paddington,* "The Herb of Death" throws the reader a curve based on our limited perspective.[128] Through the limited perspective of the reader on the garden, we learn things are not as alive as we think. The curve in "The Herb of Death" advances a coming modernity that shuts down the vibrant field of women's knowledge of plants gained during the war just as they were historically seeking to expand the same. In place of the expansion of women as cultural authorities, the ending underscores the construction of a masculine medical elite who build their expertise and cultural authority by dismissing the potency of "leaves and berries" and the ecological interdependence between people and plants in modernity. The gendered ecological stakes of this struggle are much higher in Christie's later work, the Poirot mystery *Five Little Pigs,* published the year after the heated debate over the Pharmacy and Medicines Act passed in 1941. Rather than erasing the New Woman herbalist nearly entirely, this text makes a point of criminalizing her and aligning her death with the death of herbalism as a field.

"Home Brewing": Five Little Pigs

Five Little Pigs, also titled *Murder in Retrospect,* views herbalism from two distinct historical moments in the fortunes of the practice: the 1920s and the 1940s. The novel opens in the latter period, at the request of the now adult Caroline Crale, to revisit the earlier conviction of her mother for the murder of her father, artist Amyas Crale, by coniine, a distillation of hemlock created by a local amateur herbalist. Viewed "in retrospect" from 1942, after the passage of both the 1933 Poisons Act and the Pharmacy and Medicines Act, an amateur's possession of coniine now seems especially criminal and misguided. In the course of the novel the Belgian Poirot learns that in England herbs have come under increasing surveillance and regulation since the time of the murder he is reinvestigating. As Caroline Crale doubts her mother committed the crime, Poirot must look back from the present, the 1940s, to the 1920s, investigating everyone who had access to the coniine. This includes the three main female suspects, each of whom has a motive. Artist Amyas Crale has betrayed his wife Caroline with Elsa Greer, a glamorous, acquisitive, wealthy young woman whom he is painting. Young Angela, Caroline's sister, is also a suspect, as she is furious with Amyas for ordering her back to school. Each of these women is gathered in the aptly named Handcross Manor to visit their neighbor Meredith "Merry" Blake's herbarium, which contains the distillation of coniine and many other potent plant materials. The novel presents the time of the crime as an anarchic historical moment, and this text, like "The Herb of Death," is then a witness to and perhaps advocate for, as Christie may have been herself, the increasing power of the medical profession and the decreasing cultural capital of plants in the twentieth century. As the text revisits the herbalist's laboratory, it marks the beginning of the end of the herbal revival, framing regulation as a necessary element of a rational and masculinist modernity.

Taking the opposing perspective of Leyel's *The Truth about Herbs*, Christie's novel frames "modern" women's easy access to and knowledge of potent plants as a crime waiting to happen, and in retrospect it seems that men have failed to police women and plants properly in the past. Hence, while the herbalist of *Five Little Pigs*, Merry Blake, is both competent and knowledgeable, he has blithely opened the door to no less than three vengeful female characters and given them access to his laboratory and knowledge. Dismissed as an "ineffective man," "not true to type," a "Namby Pamby," or an improperly gendered man who likes "botany and butterflies," Merry Blake is no Dr. Lloyd.[129] The ultimate closing of his laboratory is made to seem common sense at novel's end primarily because of his indiscriminate sharing of all types of plant knowledge—historical and cultural, scientific and practical, modern and ancient—with women. As in "The Herb of Death," however, in *Five Little Pigs* it is women's handling of herbs in particular that contributes to the text's ultimate defoliation. As the possibility of practicing herbalism is lost at novel's end, so is the fostering of knowledge of biodiversity and the intersection between plants and people, nature and culture. The novel pushes such interdependence away from a rational present, allying "leaves and berries," as in "The Herb of Death," with an irrational femininity and a far too merry, anarchic green past signified even by the ancient medicinal plant name of the murdered man's house, the unlucky Alderbury.[130]

Physical crossings between Alderbury and Handcross Manor represent other types of crossings that need sorting in *Five Little Pigs*. The text contains "alarming[ly]" "modern" types who disobey the boundaries of convention: an artist who refuses to obey the boundaries of marriage, a feminized man who plays with plants, and four women who seem bound only by the "code of modernity."[131] The houses of the novel contribute to this theme, both physically and emotionally open to each other, separated by a narrow creek, easily crossed, even at night. Like the houses, none of the characters in the tale are "true to type"; each violates gender norms, with the women of the novel evoking standard visions of the New Woman Criminal, devoid of "decency" and with "no code" but the code of modern acquisitiveness, speed, and their own mobility to guide them.[132] Delving into the women's culture of the herbal revival gives Christie an opportunity to galvanize the alterity of these women and the households through their association with the vibrant materiality of herbs and plants themselves.

Hence, like a walking, talking *Modern Herbal*, Merry Blake narrates a plant like hemlock into life through description, naming, analysis of constituents, history, medicinal action and uses, chemical components, preparations, and dosages in the tradition of a text like Grieve's *A Modern Herbal*. In inviting women to his laboratory and giving a disquisition on hemlock, he both educates them and provides them with an opportunity to criminalize themselves. Blake explains the long history of the plant's interrelationship with human culture, including, like Grieve, a discussion of its effect on Socrates during his execution. He mentions having read widely in Old English texts, only recently reclaimed by women scholars like Agnes Arber in 1912 and Eleanour Sinclair Rohde in 1931. Like them, he is thrilled by the mystery of plants and the very idea that their potency was once a much larger presence in human life, even sacralized. Compiling all of his knowledge from his reading, Blake explains how easy it is to distill the vital property of spotted hemlock himself, producing coniine. He knows like Grieve that "every part" of the hemlock plant, "especially the fresh leaves and fruit, contains a volatile, oily alkaloid, which is so poisonous that a few drops prove fatal to a small animal."[133] Blake is like an early modern herbarium himself, an open archive of unregulated information and herbal material culture. He delights in sharing his collection and his knowledge seemingly for knowledge's sake, and, for fun, he shares it with Caroline Crale, the artist's betrayed wife; Elsa Greer, his lover; and Angela Warren, his pouting young sister-in-law.

As he relates his former hobby to Poirot in retrospect, Blake's work now seems quite dangerous. He clearly provided the murder weapon and explained its lethal potency. He recalls that the child Angela was attentive to his tutelage, returning through an open window later to steal some valerian he had let everyone sniff. Taking place before interwar legislation, Merry's herbalism makes a good case for it. As the paternalist Superintendent Hale tells Poirot near the novel's opening, the kind of "home brewing" once practiced by the novel's Merry Blake is now rightly illegal: "Coniine and AE Salts comes under

Schedule I of the Poisons Acts."[134] The aptly named Superintendent Hale's criticism of Blake's herbalism converges with an indictment of the houses' more general sexual anarchy. Blake's lack of manliness is compounded by the excessive sexual "indecency" of Elsa Greer, who is having her portrait painted in "trousers and canvas shirts and nothing else" the day Amyas Crale is poisoned.[135] No one seems in control of plants or women in the novel; in fact, plant wisdom seems to make Merry Blake "essentially an ineffective man."[136] He fails to regulate the behavior of any of the women of the novel, three of whom cross the intellectual borders or the open windows of his laboratory, stealing his knowledge and his herbs in fits of anger or despair in order to dose the unfaithful artist Amyas Crale on his last day on earth. Their lawlessness increasingly allies them with emotion over reason and with a particularly disturbing "code of modernity" that is "no code," but rather an anarchic aggression: *"Take what you want—we shall only live once!"*[137] These modern women embody the classic role of the New Woman Criminal. They have no respect for convention, "decency," paternal authority, or the institution of marriage. A shocking embodiment of the consumerist New Woman, Elsa blithely explains to her lover's wife how she will redecorate her drawing room after their divorce. Elsa's bold hypervisibility as the subject of Crale's painterly eye captures her as the eminently visible New Woman whose beauty and vitality is linked to a shallow visual culture that conceals the true character below. Elsa is dangerously uncontainable, "blazing" destructively forth in her painting, in which she is disturbingly "alive" and potent.[138] Her glamour, beauty, and acquisitiveness easily outwit the likes of poor Merry Blake.

A foil to the modernity of the novel's women, Merry Blake the herbalist is a pitiable throwback, a nearly medieval figure: "the kind of man who would devote himself readily to a romantic and honorable devotion. He would serve his lady faithfully and without hope of reward."[139] For him, "the reliving of the past has a definite attraction."[140] He quickly shares his "enthusiasm" for herbs with Poirot in the joy of recalling his former work.[141] Like the women of the herbal revival, renowned for studying and collecting Old English and early modern herbal texts, Blake shows interest in the culture of herbs, their history, literature, and complex material practices, which unite medicine and botany, nature and culture, folklore and science in fascinating ways. However, more so than any of the women of the herbal revival, Blake resists modern science, arguing "a simple decoction" makes doctors unnecessary "half the time. The French understand these things—some of their tisanes are first rate."[142]

Referring to the French, who maintained and licensed herbalists, preserving plant literacy and biodiversity while modernizing a traditional profession,

Blake "confess[es]" to Poirot that through his herbalism he had enjoyed crossing the line between reason and imagination. He recalls foraging for plants and "gather[ing] my roots at the full of the moon or whatever it was the ancients advised."[143] Blake is the ideal reader identified in reviews of *A Modern Herbal*. Both a common reader and a serious student, book knowledge is alive for him; he channels text into *materia medica*. He recalls for Poirot how he was anxious to give his guests "a special disquisition on the spotted hemlock."[144] In the tradition of the herbal revival, he has comprehensive knowledge of when to plant flowers, when to gather the fruits: "It flowers biennially. You gather the fruits when they're ripening, just before they turn yellow."[145] He also knows like Maud Grieve that it has dropped out of the British "*pharmacopoeia*—but I've proved the usefulness of it in whooping cough—and in asthma too, for that matter."[146] Blake's plant literacy takes a dangerous material turn when he shows his guests the drug itself, shrinking the distance between the ancient past and modern England, where the plant still grows freely near water, such as the stream that divides Alderbury and Handcross Manor.

Blake embodies the kind of herbalist feared and referenced in the Parliamentary debate over the Pharmacy and Medicines Act of 1941. His access to powerful plants violates the rules of modern medicine *and* plants ideas in the reckless women listening to him. Each of these characters, and herbs themselves, is continuously allied with a femininity threatening to break civilization—represented as the conventional family—apart. Even young Angela, who grows up to become an archaeologist, seems to embody the fears of the New Woman Criminal. Her bad temper has not abated in her adulthood despite her many degrees and publications; she still seems capable of eruption. Blake's interest in historical and modern herbal literature and culture as well as science and chemistry reveals his willingness to cross lines between nature and culture. However, as his knowledge is revealed to the reader and Poirot and we relive the scene in which Blake entertains a group of volatile women with the history and components of hemlock, it is his possession and indiscriminate sharing of knowledge that strikes the reader as dangerous. Crossing epistemological boundaries to regain an intimacy between people and plants gives Blake the greatest pleasure, but it gives his audience ideas through which they become women poisoners three times over. Caroline Crale, the betrayed wife, steals his coniine meaning to kill herself. Her sister, Angela, steals his valerian. Elsa Greer steals the coniine from Caroline in a fit of rage to kill her lover when she learns he means to return to his wife. With gendered herbal anarchy like this threatening to erase civilization itself, the limits imposed by modernity and the Enlightenment that subject women to men and separate medicine and botany, culture and nature, reason and the imagination, gravitas from fun, now seem quite salutary. Gendered modernity itself seems to be the poison which should be regulated out of existence.

Certainly, as Poirot probes further into the murder of Amyas Crale, the narrative represents the Crale and Blake families as the epitome of sexual anarchy, bodied forth by their immersion in the lush local ecology and by their too-open houses, doors, windows, streams, hearts, and minds. Their aestheticism is certainly something Poirot can appreciate, as he appreciates Amyas Crale's abstract art and his stunning painting of Elsa Greer. But as Poirot works, sorting through the complex intimate relationships of families, shuttling between their houses, the present, and memories of the past, he finds that each female thief embodies a type of modernity better off dead or contained; Elsa is "[in]decent" with "very modern views," Caroline Crale is a morally flexible aesthete in an open marriage, and Angela is an "alarming young woman" who never marries, becoming a prominent archaeologist and author but literally scarred by her sister's behavior.[147] As each woman steals herbs from Blake, the narrative implies that neither they nor the resident herbalist ought to be trusted with the possession of plant knowledge, even of the mild if pungent valerian. It's too risky. While Blake is described as a misguided "county gentleman . . . given to recommending quack nostrums of his own concocting," the coniine he has concocted from local spotted hemlock is deadly, feeding into the idea apparent in the 1941 legislation that herbalists are potentially dangerous even if well intentioned.[148]

Either simply unwilling or legally banned from working with any fresh herbs by the time Poirot interviews him, Blake's retirement from herbalism suggests that interdependence between plants and people more generally is risky in modernity and better left undone. The closing of his laboratory, which defoliates his life, results in a safer, but far less vibrant, modernity, one which literally paves over the vibrant presence of plants. The idea that biodiversity benefits people and vice versa is buried in the past. All evidence of plants has been removed from Meredith's house, and his herbalism now exists only in memory. The laboratory itself where the coniine was extracted now stands "abandoned," populated only by some "derelict chemical equipment" and empty shelves.[149] Once a hotbed of sexual anarchy and herbal doings, with comings and goings through windows and plant collecting by moonlight, the house and grounds of Handcross Manor have now become a sanitized youth hostel, its once green grounds paved over with "concrete."[150] The Blake house has become a hygienic version of its former self, finally running "true to type." Physical space is strictly gendered with no crossings: "Young men sleep there, girls in the house," Blake explains to Poirot during his interview.[151] These leisured, hale young human specimens now dominate the landscape. When the former herbalist shows Poirot his laboratory bereft of plants, it is clear that the latter are not triumphant in the new order. While the space is full of "ghosts," it is haunted by a distillation of Elsa Greer in the shape of her portrait. Her youth, preserved on canvas, is the only life in the room.[152]

As the tale looks back on herbalism, it casts Merry Blake's memories of his "little hobby" as nostalgia for a world of ecological entanglement better left behind. Like the women in "The Herb of Death," the feminized herbalist in *Five Little Pigs* is a dupe. He is forced to retell how he showed his guests "round—explained the various drugs to them . . . they asked about deadly nightshade and I told them about belladonna and atropine. They were very much interested."[153] Poirot thinks Blake a fool, inviting a trio of murderous modern women to steal *both* the ancient and modern knowledge and the drugs which should more properly be regulated by men. Blake blames himself for collecting "those damned drugs" and "bumbling on about them. Pleased with my little bit of knowledge. Blind, conceited fool. I pointed out that damned coniine. I even . . . read them out that passage from the *Phaedo* describing Socrates' death . . . it's haunted me ever since."[154] Now, without his herbalism, Merry Blake is as dead as the Merry England he had tried to revive. Now his laboratory contains no plants, no "potions and distillations," but only the curtained portrait of the "very modern" Elsa Greer.[155] When the curtains are drawn back and her painting is made visible, she alone in that room is vital: "Here was life. All there was, all there could be of life, of youth, of sheer blazing vitality. The face was alive and the eyes."[156]

But like Shakespeare's "prisoner" plants "pent up in walls of glass," Elsa still retains the destructive, blazing "effect" of her youth in this distillation on canvas.[157] Entombed in the herbarium, her portrait retains the powerful effect of the New Woman Criminal, her "blazing" anarchic agency, "all that there was, all that there could be of life," burning so brightly that it literally and figuratively destroys the men around her.[158] Although Elsa claims to have died herself the day she administered coniine to her lover, her anarchic, haunting image suggests the ongoing need for the containment of the New Woman in modernity. What the laboratory has ultimately distilled is a dangerous, knowing, "very modern" woman. Her crime, the fruit of stolen knowledge, makes her monstrous. Her essence is still in effect. It still exceeds that of the room's former inhabitants, herbs and plants, underscoring the risks of human/plant interdependence in the present.

This ending seems to suggest, as do some reviews of *A Modern Herbal*, that women's indiscriminate access to knowledge is dangerous and that, in the hands of women, a little knowledge, stolen from expert men, can be particularly deadly. "Abandon[ing]" knowledge of plants now seems the smart thing to do.[159] The violence inherent in the drastic measures of the 1941 ban of the use of fresh whole plants by herbalists is illuminated by the violent clashes of Christie's novel. As Poirot penetrates deeper and deeper into the past in *Five Little Pigs,* his narrative participates more and more in a historical discourse in which women, medicinal plants, and biodiversity intertwine with danger and risk. "Home brewing" is associated, in the same breath, with fast women and questionably gendered men. These examples of the female poisoner defy the stereotype of the "cozy" mystery, which gains in horror through a woman's inversion of her role in modernity, the domestic work of cooking and nurturing.[160] The irony of this novel, however, lies in a very ambitious woman gaining access to a modern version of a much older form of herbal expertise and exercising it through sexual anarchy, with "no code to restrain her."[161] Modern herbalism quickly goes not to women's heads, but to their hearts. This seems to be a form of expertise with which women cannot be trusted. All three women poisoners intertwine with feminine passion and a volatile, abstract Nature that should be regulated by men, a claim that advocates of the 1941 Medicine and Pharmacists Act would confirm.

The Gendered Ecological Fortunes of Herbs

The rise and fall of the herbal revival is a gendered struggle over who may ultimately hold potent knowledge of the interdependence of plants and people. After the 1941 Pharmacy and Medicines Act, not only did herbalism decline, but so did *knowledge* of medicinal plants, contributing to a wider defoliation of literature and culture despite the ongoing revival of domestic herb gardening. In the latter case, the stakes of herb gardening are lowered, the need to know "leaves and berries" is less a matter of life and death. Christie's representation of women poisoners in her interwar texts marks this historical shift in the gendered ecological fortunes of herbs. Between the wars, the potency of herbs and women's knowledge of the same became a challenge to dominant notions of human exceptionalism and to the completion of the triumph of science and masculinist culture over a feminized Nature. Christie's representations of women and herbs often regain knowledge of these plants as she herself had to as an apothecary's assistant between both wars. But the trajectory of her detective narratives places that knowledge in the hands of experts, rather than the average person, particularly the average woman. Her representations, however, have great value for the ecocritic, for they reveal that the defoliation of Western culture was not an inevitable side effect of modernity or industrialization. Instead, it was the result of a gendered struggle over knowledge. When read in tandem with the work of women herbalists between the wars, Christie's mysteries register the extent to which medicinal herbs did not simply vanish in the twentieth century but, rather, were contested casualties in the construction of male expertise in modernity. Reading for plants, both to advocate for them and to locate their presence in literature, may uncover debates long forgotten and reveal the extent to which stereotypes like the female poisoner and toxic plants conceal a hidden history of the pains—and pleasures—of ecological interdependence.

Finally, what remains of the herbal revival, despite the 1941 act that attempted to eradicate it, is its literature and often its objects and material culture, even its plants, which represent their historical ecology, a new system of knowledge that encouraged intimacy with, access to, and care of plants in modernity. The signs of that ecology—books, plants, and practices—have proven resilient, aided by the lifting of the 1941 ban in 1968. Recently, much of the work of the herbal revival has been digitized by the Biodiversity Heritage Library collection, in recognition of its contribution to plant life in modernity. This is certainly in keeping with the revival's goal of making their work widely accessible. The same is true of herbs, which are now ubiquitous in England domestically and are increasingly viable commercially. However, even as we regain plant literacy today, it is also important to remember the herbal revival as we reap its benefits. That such a vision of an ecological future was produced ultimately through women's needs for work, for intellectual stimulation, for credibility, for autonomy, before, during, and after the great crisis of the First World War, intervenes in our understanding of what it means to make one's mark in the Anthropocene. What the durability of the ecological stories of the women studied in this chapter and this book reveal, finally, is that distinct interests, needs, and crises may construct ecologies that benefit both human and more-than-human nature. Now more than ever we need just such mutually beneficial, strategic, local ecologies. To suppress their memory in the Anthropocene is to build an antiecological narrative that suggests that alternative ecologies did not, do not, and will not exist in modernity. That myth is of our own making, and it is countered by the vibrant ecologies of the women represented in this chapter and throughout this book.

EPILOGUE

Perhaps the happiest fact of the work studied here is its ongoing pertinence and accessibility. The ecologies of the New Woman now seem more relevant than ever, with increasing interest in the ecological legacy of the Victorians and Edwardians, in plants in general, and in the Women's Land Army. The latter finally received a national memorial in 2014.[1] New approaches to feminist ecocriticism will hopefully continue to expand our knowledge of the conceptual reworkings of Nature and the material culture of Arts and Crafts women.[2] The engagement with intensive culture displayed by women like Cockerell and Nussey and Sarah Grand now seems prophetic, with increasing interest in food justice and biodiversity; as ecologists realize that agriculture is the single largest polluter in modernity, and as awareness of food justice and food sovereignty become keen issues, the historical experiences of their generation should be required reading.

In English gardens today, herbs are ubiquitous, recognized as key plants that anchor biodiversity and that support, in their own diversity, a wide variety of pollinators. Maud Grieve's *A Modern Herbal* has yet to be replaced as the definitive text of the field. Fittingly, technology has been a friend to the New Woman and her ecologies. Northcote's, Rohde's, and Arber's herbal texts, along with at least one of Maud Grieve's original Great War–era herbal pamphlets, are digitized in the Biodiversity Heritage Library, as is, fittingly, Olive Cockerell's scientific illustration of *Oiketicus townsendi*. The journal the *Landswoman* is now digitized by the scholarly historical website "The Women's Land Army," maintained by a young historian at Cambridge University. Agatha Christie's herbal poison plots remain the most popular books in print in the world today.

The work of regaining ecological memory is begun through the circulation of such books and images, as well as through the raising of monuments, the scanning and uploading of text, and the bottling, labeling, and storing of seeds and specimens. It is not finished, however, until ecocritics connect the ecological, cultural, ideological, conceptual, and historical dots between each discrete entry, event, text, or new item in the archive so as to allow women's communities, and their shared concerns, projects, and experiences, to appear. Simply recognizing distinctive ecologies in the texts of New Women or searching for women as a category in biodiversity databases pulls the thread of shared ideas, collaborations, competition, and debate through archives and databases. The process of archival or digital cross-referencing in this case reveals cohorts, continuity from one gendered historical and social movement to another, and an ecological story otherwise invisible. In order to understand how the New Woman's literature and culture intersected with early green consciousness and the Land Question into and beyond the crisis of the Great War, it is necessary to read far beyond how each category and its literature is traditionally defined. Rather, the archive must be expanded to the study of practices, texts, objects, and images otherwise considered in isolation. The story told here is a beginning, and I hope that one last analysis of a complex material entanglement between New Women and the environment will suffice to prove just how much more work remains to be done, and how many more connections remain to be made.

I finish, then, with a New Woman's final act, a bequeathal, which is now also preserved in a major British museum. That bequeathal was a family heirloom, a ceremonial dress described as "an herbstrewer's outfit" in the Brighton and Hove Museum's textile and fashion collection and listed in their archives as "donated by Miss Helen Nussey" (see fig. 12).[3] Although displayed in a context now far from ecological, the dress contains the story of the rise, fall, and recouping of local ecologies and of the women who kept the memory of ecological history in modernity. The dress is another deliberate archaism: contemporary and historical descriptions as well as its crisp digital image on the website of the Brighton Hove Museum show it to be an extravagant Georgian rendering of a Jacobean gown. Made of cream silk, it features wild roses cascading in a garland over the bodice and more roses strewn around the hem so as to create the impression that the wearer was literally walking through flowers. That impression was important, for the dress was worn in a great procession, at the coronation of George IV in 1821, by Helen Nussey's ancestor Sarah Ann Walker (1809–1843), at the age of twelve. As the daughter of an apothecary to the royal household, Walker represented the memory of a traditional link between women, herbs, and health care, simulating the once official role of an herbstrewer as she walked behind a figure who likewise played the role of the King's herbswoman; in her place, the mature and "striking" "Miss Fellowes" led the coronation procession.[4]

The historical memory of this dress, the procession, and the royal office was a crucial one to Nussey and many of the women discussed here. The event crops up as a touchstone in their writing, where it represents the last time "common" plants like herbs were not only nationally recognized but placed "first," representing health and wellness, through the strewing of herbs at the

Figure 12. Herbstrewer's outfit. (By permission of the Royal Pavilion & Museums, Brighton & Hove)

procession.⁵ This final walk of the royal herbswoman marked a turning point in local English ecologies.⁶ In multiple texts and articles, writers like Rosalind Northcote and Alicia Amherst record the date and time of the last appearance of the herbswoman: "At ten minutes before eleven Miss Fellowes, with her six tributary herb-women heading the grand procession, appeared at the Western Gate of the Abbey."⁷ The event "was the last exercise of an ancient custom going back to days when the strewing of herbs in the king's path and in the royal apartments fulfilled a useful purpose."⁸ The dress's last appearance signified the concomitant gendered and ecological losses involved in England's embrace of monoculture and its loss of biodiversity to industrialization and laissez-faire economic agricultural policies. Although the herbstrewer's dress has most recently been displayed as an example of the height of Georgian fashion, to read the history of the dress for its importance to women like Nussey and her colleagues is to mark their persistent commitment to ecological history and, even, to each other, through their own role in the latter.

Indeed, their careful descriptions of the last walk of the royal herbstrewers explicate not just the beauty of the moment, but the cracks between the spectacle and national ecologies that the procession embodies. The use of the herbstrewers as leaders of the coronation procession itself is in fact an incoherent ideological display, a simulation that is particularly problematic from the perspective of English biodiversity. The spectacle of the procession of ersatz herbswomen has the visual effect of greening the monarch even as the economic policies of his reign were to erode the biodiversity and local alternative agricultural and herbal practices the spectacle celebrates. Nussey complains, for example, that wild roses, like the pink silk roses and foliage on the herbstrewer's dress, are among the earliest of medicinal plants that are now grown "solely for their beauty."[9] She and her cohort regain the memory of the herbswoman's official role to counter the latter, and the preservation of the dress materializes their opposition.

The dress, and its preservation and donation, may then be read not just through meanings made at the moment of the dress's production, but through meanings made at the moment of its donation, as a material reminder of historical herbswomen and of a twentieth-century community. The dress is a memento of "Miss Helen Nussey" and her lifelong preservation of English ecological culture and its contested, gendered reclamation in the twentieth century. The dress links women's ecologies across time; its transformation into inalienable national property works to memorialize both parties, bringing them into public memory.[10]

Objects like the herbstrewer's dress, like seeds in a seedbank, are crucial paths to regaining missing links, particularly when correspondence between women is unavailable, as is so often the case. A mutual interest between Rohde, Amherst, and Nussey in the dress, for example, acts as a form of substitute correspondence, revealing shared values as well as intellectual, political, and professional alliances. Their individual discussions of the dress also become acts of community in print culture, as in Eleanour Sinclair Rohde's preface to Helen Nussey's second book, *London Gardens of the Past* (1939). In their writing, both women reveal an interest in women and herbs as a historical precedent, and a point of departure, for their own modern professionalism. Multiple women point out that the official herbstrewer was a paid position—clearly a trend they might like to revive. For them, the artifact of the dress represents a precedent of professional, gendered gardening practices that were once ubiquitous in most British households. This memory of the past prompts a "vision" of the future.[11]

The latter is fully explored in Nussey's *London Gardens of the Past*, despite its title. While the book is purportedly historical, it often segues from a discussion of an ecological past into a discussion of the possibilities of an ecological future. Written in the voice of the mature "formidable and extremely capable" welfare worker she had become by 1939, Nussey narrates a firm commitment to urban ecologies, remaining an advocate for local food and a progressive agenda involving environmental justice and urban community welfare.[12] She argues for green cities in ways that have only recently come to be recognized as possible, promoting accessible fruit-bearing trees and herbs in public parks as well as roof gardens with, again, food-productive plants and bee hives. Nussey idealizes the urban physic garden as a center of community care: "We can picture the sick and the aged from the Abbey infirmary enjoying the peace and loveliness of this herb garden whilst they watched their brethren engaged on their daily toil for the good of mankind," she writes.[13] "Mulberries, vines, figs, nuts, apples, and pear" trees, she explains, "all flourished once in the City." For women who were swept up in late Victorian and early twentieth-century agricultural reform, as Nussey was, that link to the past authorized their own vision of the ideal garden often envisioned and maintained by "knowledgeable" women.[14] It was the "extravagances" and "wars" of kings and their courts that consistently pushed urban gardens and local green stuff out of reach of the average Londoner.[15] "It now requires a conscious effort to save them for him," she wrote.[16]

Both Nussey's book and her donation of the herbstrewer's dress insist that the memory of the latter is preserved to inform the future. As she transforms the dress from alienable personable property to an "inalienable" national possession, she creates a path through which the donation may now be considered an object of ecological importance like texts in the Biodiversity Heritage Library.[17] But as "Miss Helen Nussey," the giver of the dress makes a permanent demand on the archive; she becomes a "tag" in the database, as does the name of the dress ("An Herbstrewer's Outfit"). Tracking the name of the donor and the name of the dress elsewhere links to a green history shared and constructed between women and reveals a vision only now being understood as crucial to biodiversity and sustainability. The tags from donor to dress to texts reveal that the loss of biodiversity in England should be understood as a gendered process, in which the ideas and the work of women—well before the women of Greenham Common but perhaps in sympathy with them—were pushed out of historical memory.

Luckily, gifts have a "coercive" element, enacting a complex transference of agency and responsibility from the giver to the recipient.[18] Gifts demand recognition, if not reciprocity. Nussey's gift of the herbstrewer's dress to the Brighton Hove Museum, bearing her name and received in 1961 during the ascension of both monoculture and pesticides in England, may now carry forward another story. That story, told in this book of New Woman ecologies, reveals that what is counted as "ecological" is a measure of human as well as environmental welfare. In her work as an urban social worker, professional gardener, garden historian, memory keeper, and, ultimately, as a key organizer in the evacuation of children from the city to the country during the Blitz, Nussey was less nostalgic for a green utopian past than committed to a greener London in modernity as well as a more mobile and open ruralism, both of which should make room for professional, independent women such as herself and her colleagues.

Studying the donation of the herbstrewer's dress as a last act, then, is valuable because it materializes New Woman culture and the resilient plants the dress commemorates as vibrant subjects rather than as solely discursive constructs. While more-than-human nature has been studied as a symbol for the New Woman, her intersection with the former proved it to be a force unto itself. The pains and pleasures of that meeting produce ecological thought that is valuable even now. Helen Nussey again has the last word on such things, asserting confidently that "ideas survive those who give them birth."[19] This is thankfully true, but only, in this case, if ecocritics make a deliberate effort to press all sorts of women's situated experiences, histories, practices, texts, objects, and artifacts for the ecologies that they contain.

NOTES

Introduction

1. Hill to Cockerell, November 10, 1906, City of Westminster Archives, used with permission.
2. Cockerell and Nussey, *A French Garden in England*, 8, 18.
3. Hill to Cockerell, November 10, 1906.
4. Hill to Cockerell, November 10, 1906.
5. Crane, "A Garland for May Day," *Clarion*, 1895.
6. Thirsk, *Alternative Agriculture*, 147–222.
7. Taylor, "The Small Holdings Bill," 84.
8. Taylor, 84.
9. Hockin, *Two Girls on the Land*, 8.
10. Caird, "Marriage," 196.
11. Morris owned a 1636 copy of Gerard's *Herball or Generall Historie of Plants*. He used the book to trace recipes for dyes such as the red dye from the madder plant. His personal copy of the book is exhibited at the William Morris Gallery.
12. Grieve, *Culinary Herbs*, 11–12.
13. Schreiner, *Woman and Labor*, 202.
14. Gagnier, *Individualism, Decadence, and Globalization*, 61.
15. Gagnier, 62.
16. Showalter, *Sexual Anarchy*, 39.
17. Schaffer, "The New Woman," 203.
18. Gagnier, *Individualism, Decadence, and Globalization*, 61.
19. Schaffer, "The Victorian Novel and the New Woman," 731.
20. Ledger, *The New Woman*, 9.
21. Schaffer, "The Victorian Novel," 733.
22. Gagnier, *Individualism, Decadence, and Globalization*, 63.
23. Gagnier, 63.
24. Heilmann and Beetham, *New Woman Hybridities*, 10.
25. Despina Stratigakos, "Female Firsts," 64.
26. Schaffer, "The Victorian Novel and the New Woman," 743.
27. Grieve, *Modern Herbal*, 5.
28. Tuana, "Viscous Porosity," 188.
29. Tuana, 188.
30. Alaimo, *Exposed*, 6.
31. For example, influential green anarcho-communist Peter Kropotkin wrote *Mutual Aid: A Factor of Evolution* (1901), arguing that Darwin's emphasis on the "struggle for existence" in evolution rather than "mutual aid" had been greatly distorted.

32. Shaw, *Whiteway*, 31.
33. Muñoz, *Cruising Utopia*, 22.
34. Tuana, "Viscous Porosity," 188–89.
35. Bradley in Verdon, "Business and Pleasure," 401.
36. Cunningham, "He-Notes," 99.
37. Collins, "Athletic Fashion," 310.
38. *Punch*, May 26, 1894, 252.
39. Warwick, "The New Women and the Old Acres," i.
40. Grand, *Adnam's Orchard*, 7.
41. Verdon, "Business and Pleasure," 396.
42. Verdon, 393.
43. Albritton, *Green Victorians*.
44. Shiva, *Who Really Feeds the World?* 54.
45. Shiva, 43.
46. Morton, *Poetics of Spice*, 12.
47. Alaimo, "Trans-corporeal Feminism," 238.
48. Alaimo and Hekman, *Feminist Materialisms*, 6.
49. Alaimo, "Trans-Corporeal Feminism," 239.
50. Alaimo, 239; Bennett, *Vibrant Matter*, 99.
51. Caird, "Phases of Human Development," 232.
52. Caird, 232.
53. Plumwood, *Feminism and the Mastery of Nature*, 22.
54. Plumwood, 30.
55. See Gates's discussion of Spencer in *Kindred Nature*, 14–15.
56. Gates, 14–15.
57. Caird, *Phases of Human Development*, 232.
58. Caird, 232.
59. Marx, "Commodities," *Capital*, 45.
60. In a letter to Edward Carpenter, Ellis described her "inversion" or lesbianism as controversial amongst the "Higher Thought" (her New Life friends and colleagues) as "sheer purity and sweetness to me and so the best the world cd have for me." Quoted in Wallace, "Edith Ellis, Sapphic Idealism," 189. In Havelock Ellis's book *Sexual Inversion*, case 36 of Miss H. is "widely believed to be a case history of Edith"; Wallace, "Edith Ellis, Sapphic Idealism," 189.
61. King, *Bloom*, 35.
62. Grand, "The New Aspect of the Woman Question," in Nelson, 141–42.
63. King, *Bloom*, 35.
64. Caird, "A Defence of the Wild Women," 161.
65. Schreiner, *Woman and Labor*, 51.
66. Knechtel, "Olive Schreiner's Pagan Animism," 278.
67. Taylor, "The Small Holdings Bill," 12.
68. Cockerell and Nussey, *French Garden*, 15.
69. Hockin, *Two Girls*, 30.
70. Miller, *Slow Print*, 32.
71. Holmes, "Sustaining *The Earthly Paradise*," 32; Gould, *Early Green Politics*, 34.

72. Gould, *Early Green Politics*, 34.
73. Morris, *News*, 114.
74. Warwick, *William Morris*, 5.
75. Warwick, *A Woman and the War*, 101.
76. Crane, "A Garland for Mayday."
77. Turner, *Farm Production*, 211.
78. Turner, 211.
79. Thirsk, *Alternative Agriculture*, 149.
80. Lady Warwick, *Coventry Herald*, July 21, 1899.
81. The Pankhurst family, for example, sent two children to horticultural programs. See Coleman, *Wayward Suffragette: Adela Pankhurst 1855–1961*, 52.
82. Stead presented a copy of the book to King Edward VII; see "Presentation of *French Garden in England* to the King," *Times*, March 11, 1909. A showman, Stead then published the thank-you letter he received, to confirm "the interest which the King takes in French gardening" as a solution to England's agricultural depression.
83. Warwick, *Woman*, 90.
84. MacCarthy, *Simple Life*, 90.
85. Opitz, "A Triumph of Brains Over Brute," 31.
86. Opitz, 31.
87. Warwick, "The New Women and the Old Acres," 1.
88. Warwick, *Woman*, 85.
89. Warwick, 85–86.
90. Warwick, 85.
91. "Women and Agriculture," *Times*, 1899.
92. King, *Women Rule the Plot*, 19.
93. Mrs. Chamberlain, paper delivered at the Women's Institute, November 1900, Grosvenor Crescent London, quoted in King, *Women Rule*, 19.
94. Macaulay, *Three Days*, 33–41.
95. Warwick, *A Woman and the War*, 95.
96. UK Census Online, 1901.
97. Alaimo, "Ecology," 100.
98. The image was printed by Cockerell's brother, entomologist Theodore Dru Cockerell, in the *Entomologist*, March 1913, plate 5. Cockerell explains that it is the first image of this undocumented species, sourced from a larger body of work: a monograph of *The Bombycine Moths of North America* by A. S. Packard, which his sister had been commissioned to illustrate. The publication and illustration was delayed by Packard's illness. Cockerell returned to England before finishing the project.
99. See Kingsland, "Defining Ecology as a Science," 16.
100. Darwin, *Origin of the Species*, 79.
101. Darwin, 490, 489.
102. Eliot, *The Mill on the Floss*, 35.
103. Parkins, *Mobility and Modernity*, 49.
104. Parkins, 2.
105. Parkins, 2.
106. Ouida, *Moths*, 97.

107. Parkins, *Mobility and Modernity*, 49.

108. For example, see the following: Parkins, *Mobility and Modernity*; Menon's *Evil by Design*; Scherer's "Entomology, Fiction, Intoxication"; Slosson's Narratives of Obsession"; and Knechtel's "Olive Schreiner's Pagan Animism: An Underlying Unity."

109. Warwick, *Woman*, 94.

110. See Thirsk, *Alternative Agriculture*, 147–219.

111. Ellis, letter to Carpenter, quoted in Wallace, "Edith Ellis, Sapphic Idealism," 189; Muñoz, *Cruising Utopia*, 25.

112. Grieve, *Modern Herbal*, 1.

113. Thirsk, "Rohde."

114. Grieve, *Modern Herbal*, 7–8.

115. Sanecki, *English Herb Garden*, 78.

116. Grieve, *Modern Herbal*, 5.

117. Naeem, *Biodiversity*, 3.

118. Naeem, 3.

119. Naeem, 3.

120. Naeem, 3.

121. Haraway, "Situated Knowledges," 590.

122. Gates, *Kindred Natures*, 142, 114.

1. Elemental Ecologies

1. See Pemberton, *Out of the Shadows*, 2.

2. *Allen's Indian Mail*, 1172.

3. "Christmas is Coming!" *Punch*, December 4, 1880, 254; for recent work on De Morgan, see my own article, "The Greening of Mary De Morgan"; Pemberton, "The Fairylands of Mary De Morgan: Seedbeds of Domestic Anarchy"; Talairich-Vielmas, "The Mechanization of Feelings: Mary De Morgan's 'A Toy Princess'"; Wagner, "Seeds of Subversion in Mary De Morgan's 'The Seeds of Love'"; and Fowler, "The Golden Harp: Mary De Morgan's Centrality in Victorian Fairy-Tale Literature."

4. *Allen's Indian Mail*, 1172.

5. De Morgan, "New Trade-Unionism," 276.

6. De Morgan, 276.

7. De Morgan, 276.

8. Kropotkin, *Mutual Aid: A Factor of Evolution*, 2.

9. Cohen and Duckworth, *Elemental Ecocriticism*, 13.

10. Kropotkin, *Mutual Aid*, 2.

11. Marsh, in "Concerning Love: News from Nowhere and Gender," points out the "disappointing" tendency of Morris's utopian thought to immure women in gardens, food service, and "flower arranging" (113).

12. King, *Bloom*, 35.

13. King, 5.

14. The reference is to John Ruskin's "Of Queen's Gardens." As Weltman argues, the latter turns housewives into "rhetorically empowered queens . . . [stretching] the boundaries of domestic ideology and . . . [exploring] newly defined possibilities for women within Victorian culture." *Ruskin's Mythic Queens*, 105. But Bilston notes that Ruskin's vision is

ultimately a mythic portrait: his queen is a "queen of the air," a "Pole-star" by which women should steer," toward a perhaps incoherent vision of self-renunciation and leadership; "Queens of the Garden," 5.

15. Marsh, *Pre-Raphaelite Women*, 132.
16. King, *Bloom*, 4.
17. Seaton, *The Language of Flowers*, 19.
18. Seaton, 19.
19. Murphy, *Time Is of the Essence*, 71.
20. Morris, *Letters* 2:404.
21. Morris and Burne Jones, *Pomona* and *Flora* wall hangings (1885).
22. Murphy, *Time*, 73.
23. Delvaux, "Oh Me! Oh Me! How I Love the Earth," 139.
24. Marsh, "Utopian News," 118.
25. Morris, *Letters* 2:405.
26. Morris, *News*, 34.
27. In *Jane and May Morris: A Biographical Story, 1839–1938*, Marsh notes that De Morgan worked for the embroidery department of Morris & Co after 1885 (168). De Morgan also assisted Jane Morris with embroidering the famous coverlet, which was part of a bed set presented to William Morris in 1894 on the occasion of his sixtieth birthday; Pemberton, 145.
28. Marsh, *Jane and May Morris*, 168.
29. Callen, *Women Artists of the Arts and Crafts Movement*, 17.
30. Morris, *Letters*, 415.
31. De Morgan, "The New Trades-Unionism," 277.
32. De Morgan, "The Seeds of Love," 29.
33. De Morgan, 29.
34. De Morgan, 37.
35. De Morgan, 38.
36. De Morgan, 41.
37. Weltman, "Ruskin's Mythic Queen," 5.
38. Weltman, 39.
39. Bilston argues that in Jane Loudon's work digging is compatible with "conservative Victorian ideas about women's essential delicacy" because technology provides the Victorians "better, more efficient tools" ("Queens of the Garden," 5).
40. Miller points out the dangerous pleasures, rather than the disciplining of the New Woman criminal figure, in *Framed*, 5.
41. Miller, 5.
42. De Morgan, "The Seeds of Love," 50.
43. De Morgan, 55.
44. Fowler has argued that such endings promote "ideals of self-sacrifice in love and patience in ordeals" through which love "blooms and unblooms, nourishes and injures to the core, exalts and demands sacrifice" ("The Golden Harp," 225, 230).
45. De Morgan, "The Seeds of Love," 56.
46. Miller, *Framed*, 5.
47. De Morgan, *Windfairies*, 193.

48. The first premise of the Society for Conservation Biology is listed on their website: "There is intrinsic value in the natural diversity of organisms, the complexity of ecological systems, and the resilience created by evolutionary processes."
49. Cohen and Duckert, *Elemental Ecocriticism*, 13.
50. Cohen and Duckert, 13.
51. Cohen and Duckert, 3.
52. Taylor, "Storm-Clouds on the Horizon."
53. Taylor, *The Sky of our Manufacture*, 10.
54. Morris, "The Art of the People," 530.
55. Cohen and Duckert, *Elemental Ecocriticism*, 3.
56. Stirling, *William De Morgan and His Wife*, 79.
57. Morris, "The Art of the People," 525.
58. Stirling, *William De Morgan and his Wife*, 87.
59. Gould, *Early Green Politics*, 76.
60. Qtd. in Marsh, *Back to the Land*, 142.
61. Edward Carpenter, *Civilization*, 49.
62. Carpenter, 64.
63. Morris, "Honeysuckle," William Morris Gallery.
64. Morris, "Decorative Needlework," 193.
65. Smith, *Evelyn Pickering De Morgan*, 112.
66. Smith, 121.
67. Smith, 121.
68. Smith, 121.
69. Smith, 122.
70. Smith, 122.
71. Smith, 121.
72. Smith, 122.
73. Beetham and Heilmann, "Introduction," *New Woman Hybridities*, 3.
74. Bennett, *Vibrant Matter*, xvi.
75. Gissing, *Collected Letters*, vol. 5, 113.
76. Bennett, *Vibrant Matter*, 9.
77. De Morgan, "The Windfairies," 6.
78. De Morgan, "The Pool and the Tree," 52.
79. De Morgan, 55.
80. De Morgan, 53.
81. De Morgan, 53.
82. De Morgan, 53.
83. De Morgan, 53.
84. De Morgan, 53.
85. De Morgan, 54.
86. De Morgan, 58.
87. De Morgan.
88. Taylor, "Storm-Clouds."
89. Cohen, "The Sea Above," 114.
90. De Morgan, "The Pool and the Tree," 63.

91. Cohen, "The Sea Above," 105–33.
92. Cohen, 114.
93. De Morgan, "The Pool and the Tree," 52.
94. Cohen and Duckert, *Elemental Ecocriticism*, 5.
95. De Morgan, "The Pool and the Tree," 59.
96. Alaimo, *Exposed*, 6.
97. Mortimer-Sandilands and Erickson, *Queer Ecologies*, 37.
98. De Morgan, "The Pool and the Tree," 58.
99. Cohen, "The Sea Above," 114.
100. De Morgan, "The Pool and the Tree," 64.
101. De Morgan, 64.
102. Nixon, *Slow Violence*, 2.
103. Nixon, 2.
104. Alaimo, *Bodily Natures*, 15.
105. See Gaard, "Women, Water, Energy," 160.
106. As Shteir writes in *Cultivating Women, Cultivating Science*, "New technologies made it possible to bring more specimens home alive to display them in botanical gardens or in window and balcony gardens. . . . Botany and empire cross-fertilized at the Great Exhibition" (5).
107. Shteir, 5.
108. Hagen, *An Entangled Bank*, 11.
109. The Burne-Jones grandchildren gave De Morgan the name "Mrs. Oakchest," and it is inscribed, along with her initials, in the frontispiece of *The Windfairies*.
110. To that extent they display "an ethics that is not circumscribed by the human but is instead accountable to a material world that is never merely an external place but always the very substance of ourselves and others" (Alaimo, *Bodily Natures*, 158).
111. Society for Conservation Biology.
112. Leopold, *A Sand County Almanac*, 262.
113. De Morgan, "The Rain Maiden," 204.
114. De Morgan, 193.
115. De Morgan, 194.
116. De Morgan, 195.
117. De Morgan, 196.
118. De Morgan, 203.
119. De Morgan, 205.
120. Duncan, *Art of the Dance*, 62–63.
121. Hales. "Dancer in the Dark," 534–49.
122. DeMorgan, "The Rain Maiden," 199.
123. De Morgan, 199.
124. De Morgan, 205.
125. De Morgan, 195–201.
126. Duckert, "When It Rains," 123.
127. De Morgan, "The Rain Maiden," 206.
128. De Morgan, 207.
129. Bennett, *Vibrant Matter*, 53.

130. Bennett, 99.
131. Bennett, 208.
132. Oppermann and Iovino, "Coda," 315.
133. De Morgan, "The Windfairies," 1.
134. De Morgan, 6.
135. De Morgan, 13.
136. De Morgan, 14.
137. De Morgan, 15.
138. De Morgan, 14.
139. De Morgan, 18.
140. De Morgan, 20.
141. De Morgan, 27–29.
142. De Morgan, 30–34.
143. De Morgan, 8.
144. Wilson, *Literature and Dance*, 191.
145. Wilson, 191.
146. Duncan, *Isadora Speaks*, 33.
147. Wilson, *Literature and Dance*, 191.
148. Cohen and Duckert, *Elemental Ecologies*, 4.
149. Cohen and Duckert, 6.
150. Bennett, *Vibrant Matter*, xviii.
151. Callen, *Women of Arts and Crafts*, 17; Marsh, "May Morris," 35.
152. See Lister and Marsh, eds., *May Morris: Arts and Crafts Designer*.
153. De Morgan, "The Rain Maiden," 199.

2. "We Are Two Women"

1. MacCarthy, *Simple Life*, 90.
2. Warwick, *A Woman and the War*, 86.
3. As Capuano writes, "In *The Stones of Venice* (1851–53), John Ruskin celebrated the imprecision of manual labor as a sign of elevated humanity: 'Men were not intended to work with the accuracy of tools, to be precise and perfect in all their actions. If you will have that precision out of them, and make their fingers measure degrees like cogwheels, and their arms strike curves like compasses, you must unhumanize them'" ("On Sir Charles Bell's *The Hand*," 1833). See, as well, Miller's *Slow Print*, 32, and discussions of handicraft and manual labor in intentional communities influenced by Morris and Ruskin in MacCarthy's *The Simple Life* and Albritton and Jonsson's *Green Victorians: The Simple Life in John Ruskin's Lake District*.
4. Warwick, "New Women," i.
5. Cockerell and Nussey, *A French Garden in England*, 8.
6. Warwick estimates the English imported up to 70 percent of their food at the onset of the Great War.
7. Cockerell and Nussey, *A French Garden*, 8.
8. Cockerell and Nussey, 14.
9. Thirsk, *Alternative Agriculture*, 147–222.
10. Thirsk, 148.

11. Thirsk estimates that the poor ate vegetables only once a week. It was not profitable for large farms or large markets to pay to transport a vegetable like cabbage to market. Thirsk, 148.
12. See Nilsen, *The Working Man's Green Space.*
13. McKay, 11.
14. McKay, *The French Garden,* 6.
15. McKay, 7, 13.
16. McKay, 13–14.
17. Thomas, iv.
18. Opitz, "Back to the Land," 119.
19. Bilston, Cockerell and Nussey, 14.
20. McKay, *The French Garden,* frontispiece.
21. Matthew Beaumont, "Feminism and Utopianism," 222; Rita Felski, Gender of Modernity, 47.
22. Beaumont, 215.
23. Beaumont, 214.
24. Beaumont, 214.
25. Beaumont, 216.
26. Cockerell and Nussey, *A French Garden,* 14.
27. Cockerell and Nussey, 13.
28. Cockerell and Nussey, 13.
29. Cockerell and Nussey, 14.
30. Cockerell and Nussey, 14.
31. Szerszynski, "The Post-Ecological Condition," 351.
32. Szerszynski, 351.
33. Cockerell and Nussey, *A French Garden,* 14.
34. Cockerell and Nussey, 7.
35. Coleman, *The Winter Harvest Handbook,* 236; Thirsk, *Alternative Agriculture,* 301n.
36. David Tilman, "Global Food Demand and the Sustainable Intensification of Agriculture," 20260–64.
37. Murphy, *Time Is of the Essence,* 71.
38. Cockerell and Nussey, *A French Garden,* 14.
39. The authors quote a Hungarian folk ballad, "The Lute-player's House," from *The Bard of the Dimbovitza,* 1902.
40. Cockerell and Nussey, *A French Garden,* 28.
41. Cockerell and Nussey, 28–29.
42. Cockerell and Nussey, 9.
43. Cockerell and Nussey, 24.
44. Woolf, *A Room of One's Own,* 29.
45. For example, Nussey expresses some consternation that her father has asked her several times to adjust the lease with her landlord. She writes regretting that she must "trouble him again" to make further changes at her father's request and states: "I think he is anxious lest we should be letting ourselves in for heavy expenditures in the future as he does not quite realize what a bungalow is." Nussey, Letter to Wilfred Blunt, December 3, 1907, West Sussex Record Office.

46. Collins, "Did Mid-Victorian Agriculture Fail?" 8.
47. Hockin, *Two Girls on the Land*, 112.
48. Collins, "Mid-Victorian Agriculture," 8.
49. Collins.
50. Lawrence, "The Fox," 8.
51. Lawrence, 9.
52. Lawrence, 18.
53. Lawrence, 17.
54. Lawrence, 15.
55. Lawrence, 24.
56. Lawrence, 69.
57. Lawrence, 70.
58. Cockerell and Nussey, *French Garden*, 137.
59. Thirsk, *Alternative Agriculture*, 149.
60. Morris, *News*, 228.
61. Cockerell and Nussey, *A French Garden*, 8.
62. Cockerell and Nussey, 154.
63. Cockerell and Nussey, 51.
64. King, 42.
65. King, 154.
66. Greville, 90.
67. Kropotkin, *Agriculture*, 4.
68. Stein, "The Place Promised, That Has Not Yet Been," 286–87.
69. Stein, 286–87.
70. Bennett, *Vibrant Matter*, 9.
71. Miller, *Framed*, 32.
72. Capuano, *Changing Hands*, 11.
73. Hill, "Landlords and Tenants in London," 67.
74. Plumwood, *Feminism and the Mastery of Nature*, 30.
75. Ryley, *Making another World Possible*, 31.
76. Le Fanu, *Macaulay*, 113.
77. Le Fanu, 47–48.
78. Cockerell and Nussey, *A French Garden*, 99.
79. In fact, many vegetables had acquired French names not because they were French in origin, but because the British had stopped growing them.
80. Cockerell and Nussey, *A French Garden*, 9.
81. Cockerell and Nussey, 79.
82. Cockerell and Nussey, 99.
83. As Barbara Gates notes, women science writers' use of the fairy tale could imaginatively expand the limits of their readers' sensory perception; *Kindred Natures*, 53.
84. Cockerell and Nussey, *A French Garden*, 79.
85. Rossetti, *Christina Rossetti: Poems and Prose*, 106.
86. Watson, "Men Sell Not Such in Any Town," 61–77.
87. Cockerell and Nussey, *A French Garden*, 154.

88. Olive Cockerell to Theodore Cockerell, 1907, Theodore D.A. Cockerell Collection, Archives, University of Colorado Boulder Libraries. Cockerell Box 83-10. Used by permission.

89. Heilmann, *New Woman Strategies*, 45.

90. Ardis, 10.

91. Taylor, *Sky of Our Manufacture*, 1.

92. Nussey to Blunt, dated "Monday 1910," Blunt Collection, Box 47, West Sussex Record Office. Used by permission.

93. See Nussey's correspondence to Blunt: "The garden is in such excellent condition it will be a great pity if it has to be given up altogether" (Nussey to Blunt, August 15 1910); "Gosbrook is very crushing alone & I cannot live there, besides which I want to be with Olive" (Nussey to Blunt, July 23, 1910); "I should like very much to bring her back to Gosbrook. She loves it more than any other spot" (Nussey to Blunt, dated "Monday 1910"; all letters property of Lord Lytton, Blunt Collection, Blunt Box 47, West Sussex Record Office. Used by permission.

94. Alaimo, *Bodily Natures*, 15.

95. Nussey to Blunt, July 23, 1910. Blunt Collection Box 47, West Sussex Record Office. Used by permission.

96. Nussey to Blunt, July 23, 1910.

97. Jane Morris to Sydney Cockerell, 1910, *Collected Letters of Jane Morris*, 440.

98. Alaimo, *Bodily Natures*, 17.

99. As Alaimo argues, there is perhaps no such thing as an "external environment. Instead, the posthuman being is entangled with the very stuff of the world. . . . The world is coextensive with themselves" ("Thinking as the Stuff of the World," 16).

100. Alaimo, 16.

101. The ending of Cockerell and Nussey's garden recalls Tuana's recommendation that "we often view nature as subdued through technology, the story of human agency affecting the natural order. But we forget to reverse the interaction . . . it behooves us to remember the viscous porosity between these phenomena" ("Viscous Porosity," 196).

102. Tuana, 196.

103. Hoyles, *Bread and Roses*, 52.

104. Nussey, *London Gardens of the Past*, 59.

105. Jane Morris to Sydney Cockerell, 1909, *Collected Letters of Jane Morris*, 420.

106. Studying it follows the example of scholars of the Victorian and Edwardian woman and nature, such as Gates, and answers recent calls to create a "literary ecology [that] may open up our perceptions and with them our understanding of our options" for ecological living. See Gates, *Kindred Natures*; and Keller, "Imagining Beyond, Beyond Imaginary," 582.

107. Nussey, *London Gardens*, 25.

108. Thirsk argues that "women have always had to use their ingenuity to infiltrate the interstices of an economy dominated by men. Sometimes their concerns have been described as trifles . . . but their activities, pioneered modestly in unnoticed places, have often proved to be a lifeline in adversity, and have then become commercial successes." Thirsk writes, "Pay attention to 'small things,' for they may in the end produce large changes" (264–66).

109. Thirsk, *Alternative Agriculture*, 264.

110. Cockerell, *The Valley of the Second Sons*, 279 (emphasis in original).

3. The New Life and the New Woman

1. Muñoz, *Cruising Utopia*, 28.
2. Wood, "What Is Sustainability Studies?" 14.
3. Crane, "A Garland for May Day."
4. Shaw, *Whiteway*, 23.
5. Ellis, *Attainment*, 309.
6. Ellis, "A Cornish Cottage Experiment," 361.
7. Buss, "Repossessing the World," 34.
8. Buss, 34; Beaumont, "Feminism and Utopia," 215.
9. Albritton, *Green Victorians*, 62.
10. Muñoz, *Cruising Utopia*, 32.
11. Dawson, "Climate Justice," 313.
12. Muñoz, *Cruising Utopia*, 32.
13. Greenway, "No Place for Women?," 202.
14. See in particular Parkins, "Edward Carpenter's Queer Ecology of the Everyday"; Adkins, "Transatlantic Dialogues in Sustainability: Edward Carpenter, Henry David Thoreau and the Literature of Simplification," in *Victorian Sustainability in Literature and Culture*; and Mayer's "Edward Carpenter, Henry Salt, and the Animal Limits of Victorian Environments," in *Victorian Writers and the Environment: Ecocritical Perspectives*.
15. See Richardson, *Love and Eugenics*, 95–131.
16. Hardy, *Alternative Nineteenth Century*, 156.
17. Blatchford, *Merrie England*, 61.
18. Shaw, *Whiteway*, 23.
19. Shaw, 24.
20. Shaw, 24.
21. Ellis, "Cornish Experiment," 361.
22. Ellis, 31.
23. Hardy, *Alternative Nineteenth Century*, 156.
24. Hardy, 156.
25. Hardy, 156.
26. Hardy, 165.
27. Carpenter, *Civilization*, 45.
28. Carpenter, 45.
29. Shaw, *Whiteway*, 19.
30. Shaw, 20.
31. Shaw, 31.
32. Shaw, 7.
33. Parkins, "Edward Carpenter Queer Ecology."
34. McKibben, "A Deeper Shade of Green," 37.
35. McKibben, 37.
36. Gagnier, 63.
37. Wallace, "Edith Ellis, Sapphic Idealism," 189.
38. Ellis, "Woman and the New Life," 5.
39. Ellis, 6.
40. Ellis, 6.

41. Ellis, 6.
42. Ellis, 6.
43. Ellis, *Attainment*, 10.
44. Ellis, 20.
45. Ellis, 24.
46. Ellis, 24.
47. Ellis, 108.
48. Ellis, 108–9.
49. Ellis, 110.
50. Ellis, 110.
51. Ellis, 116.
52. Ellis, 139.
53. Harris, "Pyromena," 29.
54. Ellis, *Attainment*, 146.
55. Ellis, 162.
56. Whitman, *Leaves of Grass*, 300.
57. Ellis, *Attainment*, 232.
58. Ellis, 232.
59. Ellis, 254.
60. Ellis, 289.
61. Ellis, 301.
62. Ellis, 315.
63. Ellis, 315.
64. Greenway, "No Place for Women?", 205.
65. Greenway argues that narrative structure, romantic imagery, and humor are distancing narrative strategies of women's utopian literature (202).
66. Ellis, *The Lover's Calendar*, frontispiece.
67. Ellis, "Cornish Cottage Experiment," 361.
68. Ellis, 363.
69. Ellis, 262.
70. Ellis, 361.
71. Ellis, 361.
72. Ellis, 361.
73. Ellis, *The Lover's Calendar*, preface.
74. Carpenter in *The Lover's Calendar*, 11.
75. Penn in *Lover's Calendar*, 28.
76. Blunt in *Lover's Calendar*, 71.
77. Macleod in *Lover's Calendar*, 96.
78. Tennyson in *Lover's Calendar*, 405.
79. Wallace, "Edith Ellis, Sapphic Idealism," 187.
80. Wilde in Ellis, *The Lover's Calendar*, 61.
81. Wallace, "Edith Ellis, Sapphic Idealism," 191.
82. Rossetti, "A Birthday," in Ellis's *Calendar*, 310.
83. H. G. Wells in *Whiteway*, opening pages.
84. Burtt, foreword to Shaw, *Whiteway*, 5.

85. Kortsch, *Dress Culture*, 141–42.
86. Shaw, *Whiteway*, 39, 109.
87. Shaw, 110.
88. Shaw, 39.
89. Shaw, 24.
90. Shaw, 25.
91. Shaw, 26.
92. Shaw, 44.
93. Shaw, 110.
94. Shaw and Andrews, the *New Order* 5 (July 1899), cited in Butler and Bundy, "George Egerton and Nellie Shaw," 27.
95. Bundy, 26.
96. Bundy, 39.
97. Wintle, "Horses, Bikes and Automobiles: New Woman on the Move," 66.
98. Rosowski, "Willa Cather's Ecology of Place," 37.
99. Bingham, *The Cotswolds: A Cultural History*, 39.
100. Shaw, *Whiteway*, 86.
101. Shaw, 44.
102. Shaw, 80.
103. Shaw, 7.
104. Shaw 82.
105. Shaw, 88.
106. Butler and Bundy, "Fact and Fiction: George Egerton and Nellie Shaw," 29.
107. See Solnit, "Mysteries of Thoreau Unsolved: On the Dirtiness of Laundry and the Strength of Sisters," 18–23.
108. Shaw, *Whiteway*, 88–89.
109. Burtt, foreword to Shaw, *Whiteway*, 5.
110. Shaw, *Whiteway*, 220
111. "National Character Area 107," Cotswolds, Natural England Publications. http://publications.naturalengland.org.uk/publication/5900626.
112. Butler and Bundy, "Fact and Fiction," 32.
113. Shaw, *Whiteway*, 7.
114. D'hoker, "Half-Man or Half-Doll: George Egerton's Response to Friedrich Nietzsche," 524.
115. Egerton, *Discords*, 202.
116. Egerton, 202–3.
117. Egerton, 203.
118. Egerton, 203.
119. Egerton, 203.
120. Cohen, "The Sea Above," in Cohen and Duckert, 107.
121. D'hoker, 33.
122. Egerton, *Discords*, 208.
123. Egerton, 219.
124. Egerton, 232.
125. Egerton, 211.

126. Egerton, 229.
127. Muñoz, *Cruising Utopia*, 171.
128. Muñoz, 225.
129. Muñoz, 253.
130. Hager, "A Community of Women," 26.
131. Hager, 26.
132. Warwick, "The New Women and the Old Acres," i.

133. Prendiville writes that Charles Rothschild (1877–1923), for example, drew up a list of "nature sites," which he submitted to the government of the day as worthy of state protection, and that the link between the NT and government was almost organic, with Brookes and Richardson referring to it as a "special relationship."

134. Grand, *Adnam's Orchard*, 586.

135. In *Love and Eugenics in the Late Nineteenth Century*, Richardson writes that to place Grand in "a tradition of radical nineteenth-century thought on the industrialized city and its discontents, from William Wordsworth and William Cobbett through Elizabeth Gaskell to Karl Marx, William Morris, and Edward Carpenter, or to interpret her work in the light of the environmentalist 'Back to Nature' movement of the late nineteenth century, would be to overlook the idea of racial improvement which motivated her" (141). However, it would be overly optimistic to consider any list, even of "radical" nineteenth-century protoenvironmentalists, apart from troubling strains of nationalism, classism, or racism such as the type Grand demonstrates. Edward Carpenter, for example, was strongly eugenicist and potentially racist, as were his close associates Havelock Ellis and Edith Havelock Ellis, an admirer of Morris and the Back to the Land ethos. As neo-pagan Olive Hockin shows in her writings, the "Back to Nature" movement often dovetailed easily with the politics of eugenics. Indeed, the link between early environmentalism, conservation, and eugenics has long been established, producing concerns from an environmental justice perspective about which people and places are considered worthy of national protection or conservation in both the United States and Britain. See Waterton, *Politics, Policy, and the Discourse of Heritage in Britain*.

136. Ruskin, *On the Nature of the Gothic*, 151.

137. Waterton notes that heritage policy in Britain is "held captive" to "selective understandings of a good, grand and monumental past 'owned' and monopolized by the white upper and middle classes." This discourse strongly resembles eugenicist principles in play when organizations like the National Trust were formed. *Politics*, 72.

138. Darré, quoted in Bauman, *Modernity and the Holocaust*, 113.

139. Nixon speaks of the representational bias toward a "spectacular, immediately sensational, and instantly hypervisible image of what constitutes a violent threat." If the spectacle of 9/11 has become "the definitive image of violence," the same representational bias sets "back by years attempts to rally public sentiment against climate change, a threat that is incremental, exponential and far less sensationally visible." This invisibility extends to human casualties of climate change or environmental damage as well, who often become invisible in modernity. *Slow Violence*, 13.

140. Grand, *Adnam's Orchard*, 11
141. Grand, 12.
142. Helmreich, *English Garden*, 44.
143. Helmreich, 44.

144. Helmreich, 44.
145. Grand, *Adnam's Orchard*, 454–55.
146. Helmreich, *English Garden*, 5.
147. Helmreich, 8.
148. This trend had been underway since the repeal of the Corn Laws. Imports of wheat shifted from 7 to 8 percent, to $^1/_5$ percent on repeal and ¼ percent in the 1850s to 40 percent of the total in the 1860s (Turner, *Farm Production*, 224). Still, farmers did not fully feel "the cold winds of change which overtook [the wheat market] in the 1880s, particularly the influx of cheap grain from North America. As it was, the unprotected grain farmer was caught cold. Rents collapsed, and growth fell back" (224–25). This entered in the "great crisis of the late nineteenth century agricultural depression" (Turner, 211).
149. Helmreich, *English Garden*, 27.
150. Grand, *Adnam's Orchard*, 375.
151. Grand, "The New Aspect of the Woman Question," 144.
152. Grand, 29–31.
153. Grand, 156.
154. Grand, 155.
155. Grand, 109.
156. Grand, 109.
157. Grand, 110.
158. Grand, 117.
159. Grand, 110.
160. Schaffer, "Nothing but Foolscap and Ink," 42.
161. Schaffer, 40.
162. Schaffer, 40.
163. Schaffer notes that when both Ouida and Grand deploy mythic iconography language in their debates over the New Woman, their language becomes violent, turbulent, and "volcanic" (42).
164. Schaffer, 573.
165. Legler, "Ecofeminist Literary Criticism," 230.
166. Grand, *Adnam's Orchard*, 160–70.
167. Grand, 588.
168. Grand, 598.
169. Nixon, *Slow Violence*, 13.
170. Grand, *Adnam's Orchard*, 586.
171. "Review of *Adnam's Orchard*," *Observer*.
172. "Review of *Adnam's Orchard*."
173. See Uekoetter on the "diverse roots" of Nazi ecologies in *The Green and the Brown*. See also Cronon's discussion of exclusive ecologies in "The Trouble with Wilderness."

4. "God Speed the Plough and the Woman Who Drives It"

1. Warwick, *A Woman and the War*, 69.
2. Emmeline Pankhurst played a role by helping organize a "Woman's Right to Serve" march, which included agricultural work amongst its women's wartime professions in July 1915. *Daily Express*, July 17, 1915.

3. Schreiner, *Woman and Labor,* 195.

4. See Sackville-West, *The Women's Land Army,* 15–38.

5. Understanding that gendered practices like monoculture on the English farm prevent England and interest groups like middle-class women from growing alternatives is an example of how tracing where one's food comes from reveals "political and ethical interests usually seen as separate" (Alaimo, *Material Feminisms,* 259).

6. Hockin, *Two Girls on the Land,* 8.

7. "To the Members of the Women's Land Army" address, Bedfordshire Archives.

8. Grayzel notes the discursive nostalgic tone of popular and official discourse on the project of the Women's Land Army: "Even while [the Land Army] prepared the groundwork for women transgressing sartorial, occupational, and geographic boundaries, it relocated *the* site of feminine virtue in the countryside" ("Nostalgia, Gender and the Countryside" 168). While this nostalgic discourse is certainly active in much Land Army propaganda and often in the popular press, it is often overturned in the written work published by women land workers themselves. Their experience is both material and discursive and often dialectical. What remains to be read now, especially as we consider the role of agriculture in modernity, are the ecological and political critiques landswomen share, their insights into land work, and their reflections on alternatives, many of which question how food is to be grown more sustainably in modernity. In their own words Landswomen express these concerns, which are far from nostalgic.

9. See Collingham's "How the Empire Impacted on Subsistence Farming in East Africa and Introduced Colonial Malnutrition," in *The Taste of Empire,* 239–48.

10. As Shiva points out in her 2016 introduction to *Water Wars: Privatization, Pollution, and Profit,* industrialized agriculture currently "destroys the water-holding capacity of soil. . . . Chemical agriculture . . . has also mined fertility and contributed, in great part to climate change" (xxiii).

11. The Judeo-Christian stewardship model of environmental care has been criticized as taking a managerial role. As people manage the earth in this model, it regards the world as a resource to be managed for their benefit. This characterizes the vitality of more-than-human nature as existing in an ordained service to humanity. Some have claimed it is "an arrogant ethic" for this managerial assumption and that it places "humans in a unique and privileged position." See Palmer, "Stewardship: A Case Study in Environmental Ethic," 63–75.

12. The effect of an occupying force was institutionalized through the Defense of the Realm Act (DORA). This act gave the government power to take over land and, in 1917, under the Cultivation of Lands Order, War Agricultural Executive Committees were formed in each county to help increase food production in Britain. They enforced the orders to "Plough up Pasture for Crop Planting" that aimed to bring fields used for grazing into arable production and saw over 2.5 million acres of grazing land ploughed. The campaign was not as successful as hoped; farmers had no choice on which fields they had to sow a particular crop, and many farmers only had experience of livestock.

13. White, "Feeding the War Effort," 100.

14. Hockin, *Two Girls,* 83.

15. As White argues, patriotic and sentimental cultural construction of the memory of the Land Army has effectively obscured the original intentions of the organizers and in

particular the alliance between women's land work and the values of the Back to the Land movement promoted by the director of the Women's Land Army, Meriel Talbot. White, 7, 5.

16. White, *The Women's Land Army in First World War Britain*, 3.

17. As White explains the condition of the fields, "By abandoning their normal crop rotations the land soon became weedy and infertile and instead of improving the productivity of the country, the move resulted in financial losses and a drop in agricultural production" (13).

18. McMahon, "Resisting Globalization: Women Organic Farmers and Local Food Systems," 204.

19. Hockin, *Two Girls on the Land*, 70.

20. Sackville-West, *The Women's Land Army*, 9.

21. White, *The Women's Land Army*, 3.

22. Cohen, *Remapping*, 50.

23. Bennett, *Vibrant Matter*, 9.

24. Gowdy-Wygant, *Cultivating Victory*, 15–16.

25. Warwick, *A Woman and the War*, 101.

26. Warwick, 101.

27. Warwick, 17.

28. Warwick, 17.

29. Warwick, 17.

30. Warwick, 17.

31. Warwick, 69.

32. Warwick, 69.

33. Warwick, 49.

34. Warwick, 38.

35. Wilkins, *Training and Employment of Educated Women*, i.

36. Louisa Wilkins, letter to the *Times*, June 1, 1916.

37. Wilkins.

38. Otto and Rocco, *The New Woman International*, 3.

39. Otto and Rocco, 3.

40. Talbot, "Foreword," 1.

41. Talbot, 1.

42. Hockin, *Two Girls on the Land*, iv.

43. "The Recruiter," *Landswoman*, October 1918, 217.

44. "Imagination/Realization," *Landswoman*, September 1918, 190.

45. "Imagination/Relization," 284.

46. "Molly on the Land, The Land on Molly," the *Landswoman*, April 1919, 86.

47. Gamages advertisement, *Landswoman*.

48. Oatine advertisement, *Landswoman*, January 1918, cover page.

49. Jerry and Jubbs, *Landswoman*, November 1918, 254.

50. Mulvey, "Visual Pleasure and Narrative Cinema," 843.

51. Plumwood argues for a critique of reason familiar to feminist and postmodern thought, a challenging of Enlightenment dualisms that gender the nature/culture divide. She argues as well for the expulsion of the "master" model, which argues "civilization" is marked by its transcendence of need or necessity. As Plumwood says, men engaged with

the master model resist "the recognition of dependence but continue to conceptually order [their] world in terms of a male (and truly human) sphere of free activity taking place against a female (and natural) background of necessity" (*Feminism and the Mastery of Nature*, 22). As Alaimo and Hekman argue, materialist feminists are now engaged in the project of "radically rethink[ing] materiality, the very 'stuff' of bodies and natures" (*Material Feminisms*, 6).

52. McMahon, "Resisting Globalization: Women Organic Farmers and Local Food Systems," 204.

53. R. E. B, "The Farmyard on Strike," *Landswoman,* April 1919, 86.

54. Cohen, *Remapping,* 3–10.

55. Van Puymbroeck, "Becoming a Land Girl," 398.

56. Talbot, editor's letter to the Landswomen, *Landswoman,* August, 1918, 172.

57. Cunningham, "He-Notes: Reconstructing Masculinity," in *The New Woman in Fiction and Fact,* 99.

58. Hockin, *Two Girls on the Land,* 157.

59. Hockin served four months in Holloway House prison for bombing the Roehampton Golf Club in 1913. She was also suspected of involvement in damaging the orchid house at Kew and the firebombing of Prime Minister Lloyd George's house; however, the evidence was not sufficient to convict her. There was significant bomb-making equipment found in her apartment after the Golf Club bombing and a newspaper with her name on it was found at the scene at Roehampton.

60. Ouditt's reading of Hockin's memoir critiques her "enraptured" response to biodiversity in Devonshire and her general "ignorance and sentimentality" regarding farm and country life as character flaws and signs of her superior class position, which prevent Hockin from rationalizing the need for brutal practices on the farm (*Fighting Forces, Writing Women,* 58–60). In overstating class opposition, the "impoverishment" of the middle-class farmers for whom Hockin works and their conflict, Ouditt does not acknowledge the value of Hockin's critique of gendered and classed divisions of labor on the farm, all of which are in service to the primary beneficiary, the farmer and his wife, despite their heavy cost to laborers and the land itself (60).

61. Hockin, *Two Girls on the Land,* 88.

62. Felski, *The Gender of Modernity,* 47.

63. See Marsh's discussion of this movement in *Back to the Land.*

64. See Gould on the role of mysticism in the Back to Nature movement in *Early Green Politics,* 45–57.

65. Hockin, *Two Girls,* 88.

66. Hockin, 88.

67. Hockin, 89.

68. Hockin, 22.

69. Hockin, 13.

70. Heilmann and Beetham, *New Woman Hybridities,* 10.

71. Talbot, editor's letter to the Landswomen, 172.

72. Bennett, *Vibrant Matter,* 104; Hockin, 88.

73. Hockin, 19.

74. Beaumont sees the tradition of much of the New Woman's utopian writing as a wider "call to arms," a building of a "new society... base[d] on the political bonds forged between writer, reader and a wider audience ("Feminism and Utopianism at the Fin de Siècle," 222).
75. Heilmann and Beetham, *New Woman Hybridities*, 10.
76. Talbot, 172.
77. Hockin, 154.
78. Hockin, 155.
79. Hockin, 155.
80. Hockin, 156.
81. Talbot, 172.
82. Talbot, 172.
83. Talbot, 172.
84. Talbot, 88.
85. Talbot, 157.
86. Talbot, 157.
87. Talbot, 157–58.
88. Smith, "Macaulay."
89. Le Fanu, 113.
90. Le Fanu, 113.
91. Grayzel, 35.
92. Postpastoral literature, Gifford argues, asks the following six questions: "Can awe in the face of natural phenomena such as landscapes lead to humility in our species?.... What are the implications of recognizing that we are part of a creative-destructive process?.... If the processes of our inner nature echo those in outer nature in the ebbs and flows of growth and decay, how can we learn to understand the inner by being closer to the outer?.... If nature is culture, is culture nature?.... Can our distinctively human consciousness ... be used as a tool to heal our troubled relationship with our natural home?.... How should we address the issue that the exploitation of our planet emerges from the same mind-set as our exploitation of each other?" (*Reconecting with John Muir: Essays in Post-Pastoral Practice*, 31–35).
93. Stein, "The Place Promised, That Has Not Yet Been," 286–87.
94. Macaulay, "Dedicatory," *Three Days*.
95. Macaulay, *Three Days*, 41.
96. Macaulay, 41.
97. Macaulay, 33.
98. Alpers, *What is Pastoral?*, 187.
99. Alpers, 187.
100. Macaulay, "Driving Sheep," 28.
101. Macaulay, 28.
102. Macaulay, 29.
103. Macaulay, 29.
104. Macaulay, 28.
105. Grieve, *A Modern Herbal*, 370.
106. Macaulay, "Burning Twitch," 29.
107. Macaulay, 29.

108. Macaulay, 30.
109. Macaulay, 30.
110. Gifford, "Pastoral, Anti-Pastoral, and Post-Pastoral as Reading Strategies," 42.
111. Bentley, *Eating for Victory,* 98.
112. Macaulay, *Three Days,* 31.
113. Macaulay, 32.
114. Macaulay, 32.
115. Macaulay, 32.
116. Macaulay, 33.
117. Macaulay, 33.
118. Macaulay, 33
119. Macaulay, 33.
120. Macaulay, 32.
121. Macaulay, 32.
122. Macaulay, 34.
123. Macaulay, 32.
124. Macaulay, 34.
125. Macaulay, 34.
126. Macaulay, 34.
127. Macaulay, 35.
128. Macaulay, 35.
129. Macaulay, 35.
130. Yaeger,"*Beasts of the Southern Wild* and Dirty Ecology."
131. Yaeger.
132. Yaeger.
133. Yaeger.
134. Yaeger.
135. Yaeger.
136. Le Fanu, *Rose Macaulay,* 113.
137. Talbot, "A Message to the Land Army from Miss Talbot," 103.
138. See White, *The Women's Land Army in First World War Britain,* 131–54.
139. Editor, the *Landswoman,* November 1919, 256.
140. King, *Women Rule the Plot,* 101.
141. The first issue of the *Land Girl* was published April 1, 1940.
142. Meredith, "From Ideals to Reality: The Women's Smallholding Colony at Lingfield, 1920–1939," 105.
143. Department for Environment, Food, and Rural Affairs, "Food Statistics," January 2018.
144. "Environmental Tipping Points and Food System Dynamics: Executive Summary," 2017, 7.
145. "Environmental Tipping Points," 7.

5. Working Relationships

1. Grieve, *Culinary Herbs,* 1.
2. Grieve, 5.

3. Grieve, 8.
4. Grieve, 11.
5. Sarnecki, *English Herb Garden*, 88.
6. Arber, *Herbals*, 5.
7. Holmes, "Medicinal Herbs," 35.
8. Sarnecki, *English Herb Garden*, 80.
9. See review of *A Modern Herbal*, *Times London Literary Supplement*, August 6, 1931, 605.
10. Bennett, *Vibrant Matter*, 10.
11. Leyel, introduction to *A Modern Herbal*, xiii.
12. Anonymous review, "*A Modern Herbal*," 1931, 6.
13. Anonymous, 87.
14. Anonymous, 85.
15. Miller.
16. Christie, *Five Little Pigs*, 163.
17. Christie, 110.
18. Thirsk, *Alternative Agriculture*, 188–89; Thirsk, "Rohde."
19. Rohde, *Old English Herbals*, 2.
20. Grieve, *A Modern Herbal*, subtitle.
21. Todd, "This Speaking Garden" (review of *A Modern Herbal*), 1931, 25.
22. John Lane quoted in J. May, *John Lane and the Nineties*, 86.
23. Max Beerbohm, quoted in May, *John Lane and the Nineties*, 155.
24. *Spectator*, 745.
25. Schaffer's discussion of "The Seed of the Sun" in *The Forgotten Female Aesthetes*, 54–56, links the seed to a Gothic representation of maternity.
26. Northcote was an advocate of the preservation of her native countryside and animal rights.
27. Northcote, *Book of Herbs*, 1.
28. Northcote, 1.
29. Knight, *Of Books and Botany*, 100.
30. Northcote, 1–2.
31. Northcote, 121.
32. Northcote, 5–6.
33. Schaffer, *The Forgotten Female Aesthetes*, 86.
34. Shakespeare, Sonnet 5, in Northcote, 119.
35. Gerarde, quoted in Northcote, 163.
36. Schaffer, *The Forgotten Female Aesthetes*, 86.
37. Rohde, *A Garden of Herbs*, 4.
38. Arber, *Herbals*, 5.
39. Arber, 5.
40. Arber, 5.
41. "In the words of Henry Lyte, the translator of Dodoens," she writes, ""I thinke it sufficient for any, whom reason may satisfie, by way of answeare to alleage this action and sententious position: *Bonum, quo communius, eo melius et præstantius*: a good thing the more common it is, the better it is'" (5).
42. Arber, 5.

43. Arber, 157.
44. Arber, 157.
45. Arber, 157.
46. Arber, 223.
47. Arber, 224.
48. Arber, 224.
49. Arber, 224.
50. Edwards, "Tom Hart Dyke, plant hunter and world gardener, Lullingston Castle"; Thirsk, "Eleanour Sinclair Rohde," *Oxford Dictionary of National Biography.*
51. Rohde, *Old English Herbals,* 2.
52. Rohde, 3–4.
53. Rohde, 13.
54. Rohde, 13.
55. Rohde, 13.
56. Rohde, 4.
57. Rohde, *A Garden of Herbs,* 4.
58. Rohde, *Old English Herbals,* 9.
59. Ruskin in Rohde, *A Garden of Herbs,* 1.
60. Rohde, *A Garden of Herbs,* 1.
61. See Laith's discussion of the defoliation of modern literature and culture in his introduction to *Critical Plant Studies.*
62. Eliot, "The Waste Land," ll. 19–20.
63. Eliot, l.40.
64. Rohde, *Garden of Herbs,* 7.
65. Rohde, 1.
66. Sarnecki, *History of the English Herb Garden,* 86.
67. Grieve, *Culinary Herbs,* 11–12.
68. Grieve, 11–12.
69. Grieve, 11–12.
70. As Merchant argues, "The removal of animistic, organic assumptions about the cosmos constituted the death of nature—the most far-reaching effect of the Scientific Revolution. Because nature was now viewed as a system of dead, inert particles moved by external, rather than inherent forces, the mechanical framework itself could legitimate the manipulation of nature. Moreover, as a conceptual framework, the mechanical order had associated with it a framework of values based on power, fully compatible with the directions taken by commercial capitalism" (*The Death of Nature,* 193).
71. Merchant, 193.
72. Sanecki, *History,* 100.
73. Grieve, *A Modern Herbal,* 546, 582.
74. Grieve, *Culinary Herbs,* 4.
75. Fowles, review of the second edition of *A Modern Herbal* in the *New Statesman.*
76. On the prevalence of the perception of modernity and Nature's "fatal" intersection, see Ursula Heise's *Imagining Extinction,* 1–18.
77. Grieve, *Culinary Herbs,* 1.
78. Grieve, 7.

79. Grieve, 12.

80. Heise notes that only 4 percent of plants have been evaluated by the IUCN Red List criteria. "The selection of species that have been assessed," she writes, "is subject to an obvious 'taxonomic bias' that privileges certain kinds of species and disfavors others." She notes that the IUCN itself states that "assessments of plants, fungi, and invertebrates need to be substantially increased to represent the diversity of life adequately" (76).

81. Bennett, *Vibrant Matter*, 31

82. Bennett, 32.

83. Grieve, *A Modern Herbal*, 323.

84. Grieve, 323.

85. Grieve, 323.

86. Grieve, 323.

87. Native plants are four times more attractive than exotic imports to native bees.

88. Bennett, 6.

89. Bennett, 101.

90. Grieve, *A Modern Herbal*, 397.

91. Grieve, 397.

92. Grieve, 404.

93. Grieve, 467.

94. Grieve, 467.

95. Grieve, 467.

96. Grieve, 470.

97. Grieve, 467.

98. Grieve, 469.

99. Grieve, 470.

100. Grieve, 471.

101. *Down to Earth Women*, 72.

102. Todd, "This Speaking Garden," 25.

103. Hill, Parliamentary debate, 1941.

104. Hill.

105. Thirsk, *Alternative Agriculture*, 188–89.

106. Leyel, *The Truth about Herbs*, 88.

107. Leyel, 28.

108. Leyel, 31.

109. Leyel, 31.

110. Leyel, 28.

111. Christie, *Five Little Pigs*, 27.

112. Christie, "The Herb of Death," 180.

113. Christie, *Five Little Pigs*, 110.

114. Miller, *Framed*, 3.

115. Christie, *Five Little Pigs*, 27.

116. Christie, "The Herb of Death," 191.

117. Christie, 181.

118. Christie, 181, 180.

119. Christie, 181.
120. Christie, 175.
121. Grieve, *A Modern Herbal*, 324.
122. Christie, "The Herb of Death," 181.
123. Christie, 175.
124. Christie, 177–78.
125. Christie, 178.
126. Christie 187.
127. Christie, 287.
128. Ewer, "Genre in Transit," 111.
129. Christie, *Five Little Pigs*, 94, 41.
130. The Alderbury tree produces a dark green dye rumored to be used by Robin Hood. It is considered an unlucky tree.
131. Christie, *Five Little Pigs*, 41.
132. Christie, 43.
133. Grieve, *A Modern Herbal*, 392.
134. Christie, *Five Little Pigs*, 57.
135. Christie, 56.
136. Christie, 94.
137. Christie, 43.
138. Christie, 43.
139. Christie, 86.
140. Christie, 89.
141. Christie, 89.
142. Christie, 90.
143. Christie, 90.
144. Christie, 90.
145. Christie, 90.
146. Christie, 90.
147. Christie, 56, 163, 22.
148. Christie, 60.
149. Christie, 94, 109.
150. Christie, 106.
151. Christie, 108.
152. Curran, *Secret Notebooks*, 133.
153. Christie, *Five Little Pigs*, 90.
154. Christie, 94–95.
155. Christie, 109, 163.
156. Christie, 110.
157. Shakespeare, Sonnet 5.
158. Christie, *Five Little Pigs*, 110.
159. Christie, 80.
160. Baučeková, "The Flavor of Murder," 44.
161. Baučeková, 43.

Epilogue

1. Examples of this include recent books such as Albritton's *Green Victorians*, Taylor's *The Sky of Our Manufacture*, and Parkins's *Victorian Sustainability in Literature and Culture*.

2. In 2018, for example, a new exhibition, "May Morris: Art and Life," opened in the William Morris Gallery.

3. "An Herbstrewer's Outfit," Brighton Hove Museum.

4. The dress was displayed in 2011 in the *Dress for Excess: Fashion in Regency England* exhibition at the Brighton Hove Museum. Curator's Chronicle, "Herbwoman and her Six Attendants," The Brighton Pavilion.

5. Northcote cites a "History of Coronation of George IV" and an article in *Nineteenth Century*, June 1902, where can be found references to "most elaborate descriptions of [the] dress, badge, mantle., etc., and also portraits of her in full attire.... There is one of Miss Fellowes and her 'maids.' She has a small basket in her left hand; from her right hand, raised high, she is letting a shower of blossoms fall." Those blossoms had been grown, Northcote points out, by the King's gardeners at Marylebone specifically for the 1821 event, suggesting that they were not already being grown locally at the time, which makes the procession an ironic celebration of herbs in what was rapidly becoming an age of importation. *Book of Herbs*, 108.

6. In keeping with a new culture of both sobriety and industrialized medicines, Queen Victoria was to cut the office of the herbstrewer.

7. Northcote, *Book of Herbs*, 108.

8. Nussey, "Walker and Nussey—Royal Apothecaries, 1784-1860," 81-89.

9. Nussey, *London Gardens*, 21.

10. I am indebted to Hoskins's work on "Agency, Biography, and Objects" for this reading of gifts (74).

11. Nussey, *London Gardens*, 59.

12. Williams et al., *The Children of London*, 88.

13. Nussey, 28-29.

14. Nussey, 61.

15. Nussey, *London Gardens*, 25.

16. Nussey, 25.

17. Hoskins, 74.

18. Hoskins. "Agency, Biography, and Objects," 74-84.

19. Nussey, *London Gardens*, 25.

BIBLIOGRAPHY

Archives Consulted
Blunt Collection, West Sussex Record Office
Theodore D. A. Cockerell Collection, Archives, University of Colorado Boulder Libraries
Berg Collection, New York Public Library
Senate House, University of London
The De Morgan Foundation
Brighton Hove Museum
City of Westminster Archives
Imperial War Museum

"1,000 Training Centres Organized." *Times*, March 22, 1917.
"A French Garden in England." *British Bee Journal* 37 (1909): 134–35.
"A French Garden in England." *Journal of the Ministry of Agriculture* 15 (1909): 976.
Alaimo, Stacey. *Bodily Natures: Science, Environment, and the Material Self*. Bloomington: Indiana University Press, 2010.
———. "Cyborg and Ecofeminist Interventions: Challenges for an Environmental Feminism." *Feminist Studies* 20, no. 1 (Spring 1994): 133–53.
———."Ecology." In *The Routledge Companion to Literature and Science*, edited by Bruce Clarke with Manuela Rossini, 105–7. New York: Routledge, 2011.
———."Trans-Corporeal Feminisms and the Ethical Space of Nature." In *Material Feminisms*, edited by Stacy Alaimo and Susan J. Kekman, 237–64. Bloomington: Indiana University Press, 2008.
———. "Thinking as the Stuff of the World." *O-Zone: A Journal of Object-Oriented Studies* 1, no. 1 (2014): 13–21.
———. 2008. *Material Feminisms*. Bloomington: Indiana University Press, 2008.
———. *Undomesticated Ground: Recasting Nature as Feminist Space*. Ithaca: Cornell University Press, 2000.
Albritton, Victoria, and Frederik Albritton Jonsson. *Green Victorians: The Simple Life in John Ruskin's Lake District*. Chicago: University of Chicago Press, 2016.
Allen, Grant. "Plain Words on the Woman Question." *Fortnightly Review* 46 (October 1889): 448–58.
Allen's Indian Mail. December 8, 1880, 1172.
Alpers, Paul. *What Is Pastoral?* Chicago: University of Chicago Press, 1996.
Anderson, Gail-Nina, and Joanne Wright. *Heaven on Earth: The Religion of Beauty in Late Victorian Art*. London: Lund Humphries, 1995.
Arber, Agnes. *Herbals, Their Origin and Evolution: A Chapter in the History of Botany, 1470–1670*. London: Cambridge University Press, 1912.

Ardis, Ann L. "Netta Syrett's Aestheticization of Everyday Life: Countering the Counter-discourse of *Aestheticism*." In *Women and British Aestheticism*, edited by Talia Schaffer and Kathy Alexis Psomiades, 233–50. Charlottesville: University of Virginia Press, 1999.

Asdal, Kristin, Tone Duglitro, and Steve Hinchcliffe. *Humans, Animals, and Biopolitics: The More-Than-Human Condition*. New York: Routledge, 2017.

Baučeková, Silvia. "The Flavor of Murder: Food and Crime in the Novels of Agatha Christie." *Prague Journal of English Studies* 1, no. 3 (2014): 35–46.

Bauman, Zygmunt. *Modernity and the Holocaust*. New York: New York University Press, 1992.

Beaumont, Matthew. "Feminism and Utopianism at the Fin de Siècle." In *The New Woman in Fiction and in Fact*, edited by Angelique Richardson and Chris Willis, 212–23. Hampshire, UK: Palgrave Macmillan, 2002.

Bell, E. M. *The Story of Hospital Almoners: The Birth of a Profession*. London: Faber & Faber, 1961.

Bender, Barbara. *Landscape, Politics, and Perspectives*. Oxford: Berg, 1993.

Bennett, Jane. *Vibrant Matter: A Political Ecology of Things*. Durham, NC: Duke University Press, 2010.

Bentley, Amy. *Eating for Victory: Food Rationing and the Politics of Domesticity*. Champaign: University of Illinois Press, 1998.

Bernstein; Steven. *The Compromise of Liberal Environmentalism*. New York: Columbia University Press, 2001.

Bilston, Sarah. "Queens of the Garden: Victorian Women Gardeners and the Rise of the Gardening Advice Text." *Victorian Literature and Culture* 36, no.1 (March 2008): 1–19.

Bingham, Jane. *The Cotswolds: A Cultural History*. New York: Oxford University Press, 2009.

Blake, William. *The Complete Poetry and Prose of William Blake*. Berkeley: University of California Press, 2008.

Blatchford, Robert. *Merrie England*. London: Clarion, 1894.

Boumehla, Penny. "Realism and the Ends of Feminism." In *Grafts: Feminist Cultural Criticism*, edited by Susan Sheridan, 323–29. London: Verso, 1988.

Buss, Helen. *Repossessing the World: Reading Memoirs by Contemporary Women*. Waterloo, CA: Wilfrid Laurier University Press, 2002.

Butler, Sharon, with Peggy and Bert Bundy. "Fact and Fiction: George Egerton and Nellie Shaw." *Feminist Review* No. 1 (1988): 25–35.

Caird, Mona. "A Defence of the So-called Wild Woman." In *The Morality of Marriage and Other Essays*, 157–91. London: George Redway, 1897.

———. "Marriage." In *A New Woman Reader*, edited by Carolyn Christensen Nelson, 185–99. Peterborough, ON: Broadview, 2001.

———. "Phases of Human Development." In *The Morality of Marriage and Other Essays*, 193–239.

Callen, Anthea. *Women Artists of the Arts and Crafts Movement, 1870–1914*. New York: Pantheon, 1979.

Capuano, Peter. *Changing Hands: Industry, Evolution, and the Reconfiguration of the Victorian Body*. Ann Arbor: University of Michigan Press, 2015.

———. "On Sir Charles Bell's *The Hand*, 1833." *BRANCH: Britain, Representation, and Nineteenth-Century History*, edited by Dino Franco Felluga. Extension of Romanticism

and Victorianism on the Web. http://www.branchcollective.org/?ps_articles=peter-capuano-on-sir-charles-bells-the-hand-1833.
Carpenter, Edward. *Civilization: Its Cause and Cures, and Other Essays*. London: George Allen & Unwin, 1921.
Carroll, Alicia. "The Greening of Mary De Morgan: The Cultivating Woman and the Ecological Imaginary in 'The Seeds of Love.'" *Victorian Review* 36, no. 2 (2010): 104–17.
Christie, Agatha. *Agatha Christie: An Autobiography*. New York: Dodd, Mead, 1977.
———. *Five Little Pigs: A Hercule Poirot Mystery*. New York: William Morrow, 2011.
———. *Miss Marple: The Complete Short Stories*. New York: William Morrow, 2011.
"Christmas Is Coming!" *Punch*, December 4, 1880, 254.
Clarke, Gill. *The Women's Land Army: A Portrait*. Bristol, UK: Sansom, 2008.
Clere, Eileen. *The Sanitary Arts: Aesthetic Culture and the Victorian Cleanliness Campaigns*. Columbus: Ohio State University Press, 2014.
Cockerell, Olive, and Helen Nussey. *A French Garden in England: A Record of the Successes and Failures of a First Year of Intensive Culture*, London: Stead's, 1909.
Cockerell, Theodore Dru. *The Valley of the Second Sons: Letters of Theodore Dru Alison Cockerell, a Young Naturalist, Writing to His Sweetheart and Her Brother About His Life in West Cliff, Westmountain Valley, Colorado, 1887–1890*. Edited by William A. Weber. Boulder, CO: Pilgrims Process, 2004.
Cohen, Deborah Rae. "Encoded Enclosures: the Wartime Novels of Stella Benson." In *The Literature of the Great War Reconsidered: Beyond Modern Memory*, edited by Patrick J. Quinn and Steven Trout, 37–54. New York: Palgrave Macmillan, 2001.
Cohen, Jeffrey Jerome. "The Sea Above." In *Elemental Ecocriticism: Thinking with Earth, Air, Water, and Fire*, edited by Jeffrey Cohen and Lowell Duckert, 105–33. Minneapolis: University of Minnesota Press, 2015.
Cohen, Jeffrey Jerome, and Lowell Duckert, eds. *Elemental Ecocriticism: Thinking with Earth, Air, Water, and Fire*. Minneapolis: University of Minnesota Press, 2015.
Cohler, Deborah. "Queer Kinship, Queer Eugenics: Edith Lees Ellis, Reproductive Futurity, and Sexual Citizenship." *Feminist Formations* 26, no. 3 (2014): 122–46.
Coleman, Eliot. *The Winter Harvest Handbook: Year-Round Vegetable Production Using Deep Organic Base and Unheated Greenhouses*. White River Junction, VT: Chelsea Green, 2009.
Coleman, Verna. *Adela Pankhurst: The Wayward Suffragette 1885–1961*. Melbourne: Melbourne University Press, 1996.
Coler, Deborah. *Citizen, Invert, Queer: Lesbianism and War in Early Twentieth-Century Britain*. Minneapolis: University of Minnesota Press, 2010.
Collins, E. J. T. "Did Mid-Victorian Agriculture Fail? Output, Productivity, and Technological Change in Nineteenth-Century Farming." *Recent Findings of Research in Economic and Social History* 21 (1995): 1–8.
Collins, Tracy J. R. "Athletic Fashion, *Punch*, and the Creation of the New Woman." *Victorian Periodicals Review* 43 no. 3 (2010): 309–35.
Coupe, Laurence, ed. *The Green Studies Reader: From Romanticism to Ecocriticism*. New York: Routledge, 2000.
Cragoe, Matthew, and Paul Readman, eds. *The Land Question in Britain, 1750–1950*. New York: Palgrave Macmillan, 2010.

Crane, Walter. "A Garland for May Day." *Clarion*, May 1, 1895.
Crawford, Elizabeth. *Enterprising Women: The Garretts and Their Circle*. London: Francis Boutle, 2009.
———. *The Women's Suffrage Movement: A Reference Guide, 1866–1928*. New York: Routledge, 2000.
Cronon, William. "The Trouble with Wilderness." In *Uncommon Ground: Rethinking the Human Place in Nature*, 69–90. New York: W. W. Norton, 1995.
Cunningham, Gail. "He-Notes: Reconstructing Masculinity." In *The New Woman in Fiction and in Fact: Fin de Siècle Feminisms*, edited by Angelique Richardson and Chris Willis, 94–106. Basingstoke, UK: Palgrave Macmillan, 2002.
Curators' Chronicle. "Herbswoman and her Six Attendants." Brighton Hove Museum.
Curran, John. *Agatha Christie's Secret Notebooks: Fifty Years of Mysteries in the Making*. New York: HarperCollins, 2009.
Darwin, Charles. *On the Origin of the Species by Means of Natural Selection*. London: John Murray, 1859.
Dawson, A. "Climate Justice: The Emerging Movement against Green Capitalism." *South Atlantic Quarterly* 109 (2010): 313–38.
Defense of the Realm Act. *London Gazette* (supplement), September 1, 1914.
De Lauretis, Teresa. *Technologies of Gender: Essays on Theory, Film, and Fiction*. Bloomington: Indiana University Press, 1987.
DeLoughrey, Elizabeth, and George B. Handley, eds. *Postcolonial Ecologies*. New York: Oxford University Press, 2011.
Delvaux, Martin. "Oh Me! Oh Me! How I Love the Earth: William Morris's *News from Nowhere* and the Birth of a Sustainable Society." *Contemporary Justice Review* 8, no. 2 (June 2005): 131–46.
D'Emilio, John. "Capitalism and Gay Identity." In *Making Trouble: Essays on Gay History, Politics, and the University*, 3–16. New York: Routledge, 1992.
De Morgan, Mary. "Co-operation in England," *Westminster Review*, May 1890, 532–540.
———. *On a Pincushion and other Fairy Tales*. New York: Garland, [1877] 1977.
———. "The New Trades-Unionism and Socialism in England." *Home-Maker*, January 1891, 336–39. Reprinted in Pemberton, *Out of the Shadows*, 272–77.
———. "A Toy Princess." In *The Broadview Anthology of Victorian Short Stories*, edited by Dennis Denisoff, 265–276. New York: Broadview, 2004.
———. *The Windfairies and Other Tales*. London: Seeley, 1900.
Department for Environment, Food, and Rural Affairs. "Food Statistics," January 2018. https://www.gov.uk/government/statistics/food-statistics-pocketbook.
D'hoker, Elke. "Half-Man or Half-Doll: George Egerton's Response to Friedrich Nietzsche." *Women's Writing* 18, no. 4 (2011): 524–46.
Diamond, Irene, and Gloria Feman Orenstein. *Reweaving the World: The Emergence of Ecofeminism*. San Francisco: Sierra Club, 1990.
Duckert, Lowell. "When it Rains." In *Material Ecocriticism*, edited by Serenella Iovino and Serpil Oppermann, 114–29. Bloomington: Indiana University Press, 2014.
Duncan, Isadora. *Isadora Speaks: Writings and Speeches of Isadora Duncan*. Edited by Franklin Rosemont. Chicago: Charles H. Kerr, 1994.
Egerton, George. "A Crossline." In *Keynotes*. London: Elkin Mathews & John Lane, 1893.

———. "Now Spring Has Come." In *Keynotes*. London: Elkin Mathews & John Lane, 1893.

———. "The Regeneration of Two." In *Discords*. London: Elkin Matthews & John Lane, 1894.

Ehneen, Sharon. *Women's Literary Collaboration, Queerness, and Late-Victorian Culture*. Burlington, VT: Ashgate, 2008.

Eliot, T. S. "The Waste Land." In *The Waste Land and Other Writings*. New York: Random House, 2002.

Elliott, B., and J. Helland, eds. *Women Artists and the Decorative Arts, 1880–1935: The Gender of Ornament*. Burlington, VT: Ashgate, 2002.

Ellis, Edith [Mrs. Havelock]. "A Cornish Experiment in Cottages." *World's Work and Play* 8 (1906): 361–63.

———. *Attainment*. London: Alston Rivers, 1909.

———. *The Lover's Calendar*. Compiled and edited by Mrs. Havelock Ellis. London: Kegan, Paul, Trench, Trübner, 1912.

"Environmental Tipping Points and Food System Dynamics: Executive Summary." Global Food Security Program, United Kingdom, 2017.

Ewers, Chris. "Genre in Transit: Agatha Christie, Trains, and the Whodunit." *Journal of Narrative Theory* 46, no. 4 (2016): 97–120.

Faithfull, Emily. "A Woman Farmer." *Victoria Magazine* 33 (1879): 534–36.

Felski, Rita. *The Gender of Modernity*. Cambridge, MA: Harvard University Press, 1995.

Fernihough, Anne A. *Freewomen and Supermen: Edwardian Radicals and Literary Modernism*. New York: Oxford University Press, 2013.

Fowler, James. "The Golden Harp: Mary De Morgan's Centrality in Victorian Fairy-Tale Literature." *Children's Literature* 33, no. 1(2005): 224–36.

Frankel, Nicholas. "The Ecology of Victorian Poetry." *Victorian Poetry* 41, no.4 (2003): 629–35.

Freedgood, Elaine. *Ideas in Things: Fugitive Meanings in the Victorian Novel*. Chicago: University of Chicago Press, 2006.

Fritz, Morgan. "'The Mesmeric Power': Sarah Grand and the Novel of the Female Orator." *Texas Studies in Literature and Language* 55, no. 4 (2013): 452–72.

Gaard, Greta. "New Directions for Ecofeminism: Toward a More Feminist Ecocriticism." *Interdisciplinary Studies in Literature and Environment (ISLE)* 17, no.4 (2010): 643–65.

———. "Women, Water, Energy: An Ecofeminist Approach." *Organization & Environment* 14, no. 2 (June 2001): 157–72.

Gagnier, Regenia. *Individualism, Decadence, and Globalization: On the Relationship of the Part to the Whole, 1859–1920*. New York: Palgrave Macmillan, 2010.

Garrard, Greg. *Teaching Ecocriticism and Green Cultural Studies*. New York: Palgrave Macmillan, 2012.

Gates, Barbara. *In Nature's Name: An Anthology of Women's Writing and Illustration, 1780–1930*. Chicago: University of Chicago Press, 2002.

———. *Kindred Natures: Victorian and Edwardian Women Embrace the Living World*. Chicago: University of Chicago Press, 1998.

Gifford, Terry. "Ecology and the Literature of the British Left: The Red and the Green." *ISLE* 20, no. 2 (Spring 2013): 444–45.

———. "Pastoral, Anti-Pastoral, and Post-Pastoral as Reading Strategies." In *Critical Insights: Nature and Environment*, edited by Scott Slovic, 42–61. Ipswich, MA: Salem Press, 2012.

———. *Reconnecting with John Muir: Essays in Post-Pastoral Practice*. Athens, GA: University of Georgia Press, 2006.

Gissing, George. *Collected Letters of George Gissing, 1892–1895*. Edited by Paul F. Mattheisenn, C. Young, and Pierre Coustillas. Athens, OH: Ohio University Press, 1994.

Goggin, Maureen Daly, and Beth Fowkes Tobin, eds. *Women and the Material Culture of Needlework and Textiles, 1750–1950*. London: Ashgate, 2009.

Gould, Peter. *Early Green Politics: Back to Nature, Back to the Land and Socialism in Britain, 1880–1900*. Brighton, UK: Harvester, 1988.

Gowdy-Wygant, Cecilia. *Cultivating Victory: The Women's Land Army and the Victory Garden Movement*. Pittsburgh: University of Pittsburgh Press, 2013.

Grand, Sarah, *Adnam's Orchard*. London: William Heinemann, 1912.

———. *The Heavenly Twins*. London: William Heinneman, 1893.

———. "The Man of the Moment." *North American Review* 158 (May 1894): 620–25. Reprinted in Nelson.

———. "The New Aspect of the Woman Question." *North American Review* 158 (March 1894): 270–76. Reprinted in Nelson.

Grayzel, Susan R. "Nostalgia, Gender, and the Countryside: Placing the 'Land Girl' in First World War Britain." *Rural History* 10, no. 2 (1999): 155–70.

Greenway, Judy. "No Place for Women? Anti-Utopianism and the Utopian Politics of the 1890s." *Geografiska Annaler: Series B, Human Geography* 84, nos. 3-4 (2002): 201–9.

Grieve, M. *A Modern Herbal: The Medicinal, Culinary, Cosmetic, and Economic Properties, Cultivation and Folk-Lore of Herbs, Grasses, Fungi, Shrubs and Trees with All Their Modern Scientific Uses*. London: Jonathan Cape, 1931.

———. *Culinary Herbs: How to Grow and Where to Sell with an Account of their Uses and History*. N.p.: British Guild of Herb Growers, n.d.

———. *Economic Trees and their By-Products*. British Guild of Herb Growers, n.d.

Grove, Richard H. *Green Imperialism: Colonial Expansion, Tropical Island Edens, and the Origins of Environmentalism, 1600–1860*. New York: Cambridge University Press, 1996.

Hagen, Joel. *An Entangled Bank: The Origins of Ecosystem Ecology*. Rutgers, NJ: Rutgers University Press, 1992

Hager, Lisa. "A Community of Women: Women's Agency and Sexuality in George Egerton's Keynotes and Discords." *Nineteenth-Century Gender Studies* 2, no. 2 (2006): http://www.ncgsjournal.com/issue22/hager.htm.

Hales, Barbara. "Dancer in the Dark: Hypnosis, Trance-Dancing, and Weimar's Fear of the New Woman." *Monatshefte* 102, no. 4 (2010): 534–49.

Hall, Sarah. *Daughters of the North*. New York: Harper Perennial, 2008.

Hansard Parliamentary Debates, 5th ser. Volume 373. United Kingdom, 1941. 63–127.

Haraway, Donna. *The Companion Species Manifesto: Dogs, People, and Significant Otherness*. Chicago: Prickly Paradigm, 2003.

———. "Situated Knowledges: The Science Question in Feminism and the Privilege of Partial Perspective." *Feminist Studies* 14, no. 3 (1988): 575–99.

Harris, Anne. "Pyromena: Fire's Doing." In *Elemental Ecocriticism*, edited by Jeffrey Jerome Cohen and Lowell Duckert, 27–54. Minneapolis: University of Minnesota Press, 2015.

Head, Dominic. "Raymond Williams and Ecocriticism." *Green Letters: Studies in Ecocriticism* 1, no. 1 (2000): 7–9.
Hearn, Michael Patrick, ed. *The Victorian Fairy Tale Book*. New York: Pantheon, 1988.
Heilmann, Ann. *New Woman Strategies: Sarah Grand, Olive Schreiner, Mona Caird*. Manchester: Manchester University Press, 2004.
Heilmann, Ann, and Margaret Beetham. *New Woman Hybridities: Femininity, Feminism, and International Consumer Culture, 1880–1930*. London: Routledge, 2004.
Heise, Ursula K. *Imagining Extinction: The Cultural Meanings of Endangered Species*. Chicago: University of Chicago Press, 2016.
Helmreich, Anne. *The English Garden and National Identity: The Competing Styles of Garden Design, 1870–1914*. New York: Cambridge University Press, 2002.
Higonnet, Margaret, Randolph Higonnet, Jane Jenson, Sonya Michel, and Margaret Collins Weitz, eds. *Behind the Lines: Gender and the Two World Wars*. New Haven: Yale University Press, 1987.
Hill, Octavia. "The Kyrle Society." *Charity Organisation Review* 18 (1905): 314–19.
———. "Landlords and Tenants in London" (1871). In *Octavia Hill and the Social Housing Debate, Essays and Letters by Octavia Hill*, edited by Robert Whelan, 67. London: Civitas, 2000.
———. *The Life of Octavia Hill as Told in Her Letters*. London: Macmillan, 1913.
———. *Our Common Land and Other Short Essays*. London: Macmillan, 1877.
Hilton, Tim. *Ruskin: The Later Years*. New Haven: Yale University Press, 2000.
Hockin, *Two Girls on the Land: War-Time on a Dartmoor Farm*. London: E. Arnold, 1918.
Holmes, E. M., FLS. "Medicinal Herbs: Their Cultivation and Preparation in Great Britain." *Journal of the International Garden Club* 2, no. 1 (1918): 35–50.
Holmes, John. "Sustaining *The Earthly Paradise*." In *Victorian Sustainability in Literature and Culture*, edited by Wendy Parkins, 32–50. London: Routledge, 2017.
Hoskins, Janet. "Agency, Biography, and Objects." In *The Handbook of Material Culture*, edited by C. Tilley et al., 74–84. London: Sage, 2006.
Houle, Karen. "Animal, Vegetable, Mineral: Ethics as Extension or Becoming? The Case of Becoming Plant." *Journal for Critical Animal Studies* 9 (2011): 89–116.
Howarth, William L. "Imagined Territory: The Writing of Wetlands." *New Literary History* 30, no. 3 (1999): 509–39.
Hoyles, Martin. *Bread and Roses: Gardening Books from 1560 to 1960*. London: Pluto, 1995.
Hughes, Linda K. "Feminizing Decadence." In *Women and British Aestheticism*, edited by Talia Schaffer and Kathy Alexis Psomiades, 119–38. Charlottesville: University of Virginia Press, 2000.
Jekyll, Gertrude. "Idea of a Garden." *Edinburgh Review*, July 1896. Reprinted in *A Gardener's Testament: A Selection of Articles and Notes*, 10–38. London: Country Life, 1937.
Jenkins, Martin. "Case Study No. 5: Effectiveness of Biodiversity Conservation." Food and Agriculture Organization of the United Nations, 2002.
Kehler, Grace. "Gertrude Jekyll and the Late-Victorian Garden Book: Representing Nature-Culture Relations." *Victorian Literature and Culture* 35, no. 2 (2007): 617–33.
Keller, Lynn. "Beyond Imagining, Imagining Beyond." *Publications of the Modern Language Association of America (PMLA)* 127, no. 3 (May 2012): 579–85.

Kerridge, Richard. "What Is Ecocriticism? UK Perspectives." *Green Letters: Studies in Ecocriticism* 1, no. 1 (2000): 4.
King, Amy M. *Bloom: The Botanical Vernacular in the English Novel.* Oxford: Oxford University Press, 2003.
King, Peter. *Women Rule the Plot: The Story of the 100 Year Fight to Establish Women's Place in Farm and Garden.* London: Duckworth, 1999.
Kingsland, Sharon. "Defining Ecology as a Science." In *Foundations of Ecology: Classic Papers with Commentaries,* 16–28. Chicago: University of Chicago Press, 1991.
Knechtel, Ruth. "Olive Schreiner's Pagan Animism: An Underlying Unity." *English Literature In Transition, 1880–1920* 53, no. 3 (2010): 259–82.
Knight, Leah. *Of Books and Botany in Early Modern England: Sixteenth-Century Plants and Print Culture.* Burlington, VT: Ashgate, 2009.
Knoepflmacher, U. C., and G. B. Tennyson, eds. *Nature and the Victorian Imagination.* Berkeley: University of California Press, 1977.
Kortsch, Christine Bayles. *Dress Culture in Late Victorian Women's Fiction: Literacy, Textiles, and Activism.* New York: Routledge, 2016.
Koven, Seth. *Slumming: Sexual and Social Politics in Victorian London.* Princeton: Princeton University Press, 2006.
Kropotkin, Prince Peter. *Mutual Aid: A Factor of Evolution.* London: William Heinemann, 1902.
———. "Proposed Communist Settlement: A New Colony for Tyneside or Wearside." *Newcastle Daily Chronicle,* February 20, 1895.
———. *The Conquest of Bread and Other Writings.* Edited by Marshall Shatz. Cambridge: Cambridge University Press, 1995.
Laist, Randy, ed. *Critical Plant Studies: Philosophy, Literature, Culture.* Amsterdam: Rodopi, 2013.
Lash, Scott, and Jonathan Friedman. eds. *Modernity and Identity.* New York: Wiley-Blackwell, 1992.
Lawrence, D. H. "The Fox." In *The Fox/The Captain's Doll/The Ladybird,* 5–72. New York: Penguin, 2006.
Ledger, Sally. *The New Woman: Fiction and Feminism at the Fin de Siècle.* Manchester: Manchester University Press, 1997.
Lefanu, Sarah. *Rose Macaulay.* London: Virago, 2003.
Legler, Gretchen. "Ecofeminist Literary Criticism." In *Ecofeminism: Women, Culture, Nature,* edited by Karren Warren and Nisvan Erkal, 227–38. Bloomington: Indiana University Press, 1997.
Leopold, Aldo. *A Sand County Almanac.* New York: Oxford University Press, 1966.
Levine, George. "High and Low: Ruskin and the Novelists." In *Nature and the Victorian Imagination,* edited by U. C. Knoepflmacher and G. B. Tennyson, 137–53. Berkeley: University of California Press, 1977.
Leyel, C. F. [Hilda]. "Introduction." *A Modern Herbal: The Medicinal, Culinary, Cosmetic and Economic Properties, Cultivation and Folklore of Herbs, Grasses, Fungi, Shrubs and Trees with All Their Modern Uses.* London: Dorset, 1992.
———. *The Truth about Herbs.* London: Andrew Dakers, 1943.

Lister, Jenny, and Jan Marsh, eds. *May Morris: Arts and Crafts Designer.* London: Thames & Hudson, 2017.
Logan, Peter. *Victorian Fetishism: Intellectuals and Primitives.* New York: State University of New York Press, 2010.
Londraville, Janice, ed. *On Poetry, Painting, and Politics: The Letters of May Morris and John Quinn.* Selingrove, PA: Susquehanna University Press, 1997.
Long, James. *Small Holdings.* Glasgow: Collins' Clear-Type Press, 1913.
Macan, Hugh. "Education Suitable for Women in Rural Districts." In *Progress in Women's Education in the British Empire,* edited by F. E. Warwick, 144–45. London: Longmans & Green, 1898.
Macaulay, Rose. *Three Days* Chiswick: Charles Wittingdon & Sons, 1919.
MacCarthy, Fiona. *The Simple Life: C. R. Ashby in the Cotswolds.* London: Faber & Faber, 1981.
MacDuffie, Allen. *Victorian Literature, Energy, and the Ecological Imagination.* New York: Cambridge University Press, 2014.
Marsh, Jan. *Back to the Land: The Pastoral Impulse in England, 1880–1914.* Ann Arbor: University of Michigan, 1982.
———. "Concerning Love: *News From Nowhere* and Gender." In *William Morris and News from Nowhere: A Vision for Our Time,* edited by Stephen Coleman and Paddy O'Sullivan, 107–25. Bideford, UK: Green, 1990.
———. "May Morris: Ubiquitous, Invisible, Arts and Craftswoman." In *Women Artists and the Decorative Arts, 1880–1935: The Gender of Ornament,* edited by B. Elliott and J. Helland, 35–52. London: Ashgate, 2002.
———. *Pre-Raphaelite Women: Images of Femininity.* London: Crown, 1988.
Marx, Karl. *Capital: A Critique of Political Economy.* Vol. I. New York: Penguin, 1992.
May, J. Lewis. *John Lane and the Nineties.* London: John Lane/The Bodley Head, 1936.
McClintock, Anne. *Imperial Leather: Race, Gender and Sexuality in the Colonial Contest.* New York: Routledge, 1995.
McKay, C. D. *The French Garden: A Diary and Manual of Intensive Cultivation.* London: Carmelite House, 1908.
McKibben, Bill. "A Deeper Shade of Green." *National Geographic* 210, no. 2 (2006): 33–37.
McMahon, Martha. "Resisting Globalization: Women Organic Farmers and Local Food Systems." *Canadian Women Studies* 21, no. 4 (Spring 2002): 203–4.
Menon, Elizabeth K. *Evil by Design: The Creation and Marketing of the Femme Fatale.* Champaign: University of Illinois Press, 2006.
Merchant, Carolyn. *The Death of Nature: Women, Ecology, and the Scientific Revolution.* New York: Harper & Row, 1990.
Meredith, Anne. "Horticultural Education in England, 1900–40: Middle-Class Women and Private Gardening Schools." *Garden History* 31, no. 1 (Spring 2003): 67–79.
———. "From Ideals to Reality: The Women's Smallholding Colony at Lingfield, 1920–1939." *Agricultural History Review* 54, no. 1 (2006): 105–21.
Miele, Chris, ed. *From William Morris: Building Conservation and the Arts and Crafts Cult of Authenticity, 1877–1939.* London: Paul Mellon Centre, 2005.

Miller, Elizabeth Carolyn. *Framed: The New Woman Criminal in British Culture at the Fin de Siècle*. Ann Arbor: University of Michigan Press, 2008.

———. "William Morris, Extraction Capitalism, and the Aesthetics of Surface." *Victorian Studies* 57, no. 3 (2015): 395–404.

———. *Slow Print: Literary Radicalism and Late Victorian Print Culture*. Stanford: Stanford University Press, 2013.

Morrell, Caroline. *Housing and the Women's Movement, 1860–1914*. PhD diss., Oxford Brookes University, 1999.

Morris, Jane. *The Collected Letters of Jane Morris*. Edited by Frank C. Sharp and Jan Marsh. Woodbridge, UK: Boydell, 2012.

Morris, May. "Decorative Needlework." In *The International Congress of Women of 1899: Women in Professions*, edited by the Countess of Aberdeen, 191–94. London: T. Fisher Unwin, 1900.

———. "William De Morgan, Recollections." *Burlington Magazine*, August and September 1917.

———. *Introductions to the Collected Works of William Morris*. 2 vols. New York: Oriole Editions, 1973.

Morris, William. *Collected Letters of William Morris*. Edited by Norman Kelvin. Princeton: Princeton University Press, 1996.

———."Hopes and Fears for Art." In *The Collected Works of William Morris: With Introductions by his Daughter May Morris*, Vol. 22, 3–119. London: Longmans, Green, 1910–15.

———. *News from Nowhere*. Edited by Clive Wilmer. New York: Penguin, 2004.

———. *William Morris: Poet, Artist, Socialist*. Edited by Frances Watts Lee. New York: Humboldt, 1891.

Morrow, Elsa. "History of Swanley Horticultural College." *Wye: The Journal of the Agricola Club and Swanley Guild* 12 (1985): 59–142.

Morton, Timothy. *Ecology without Nature: Rethinking Environmental Aesthetics*. Cambridge, MA: Harvard University Press, 2007.

———. "Guest Column: Queer Ecology." *PMLA* 125, no. 2 (2010): 273–82.

———. *The Poetics of Spice: Romantic Consumerism and the Exotic*. Cambridge: Cambridge University Press, 2000.

———. "Thinking Ecology: The Mesh, the Strange Stranger, and the Beautiful Soul." *Collapse* 6 (2010): 195–223.

"Mr. Gladstone on Cottage Gardening and Fruit Farming." *Times*, August 22, 1890.

Mulvey, Laura. "Visual Pleasure and Narrative Cinema." In *Film Theory and Criticism: Introductory Readings*, edited by Leo Braudy and Marshall Cohen, 833–44. New York: Oxford University Press, 1999.

Muñoz, José Esteban. *Cruising Utopia: The Then and There of Queer Futurity*. New York: New York University Press, 2009.

Murphy, Patricia. *Time Is of the Essence: Temporality, Gender, and the New Woman*. New York: State University of New York Press, 2001.

Naeem, Shahid, Daniel E. Bunker, Andy Hector, Michel Loreau, and Charles Perrings, eds. *Biodiversity, Ecosystem Functioning, and Human Wellbeing*. Oxford: Oxford University Press, 2009.

Nelson, Carolyn Christensen, ed. *A New Woman Reader: Fiction, Articles, and Drama of the 1890s*. Peterborough, ON: Broadview, 2002.

Nilsen, Micheline. *The Working Man's Green Space: Allotment Gardens in England, France, and Germany, 1870–1919*. Charlottesville: University of Virginia Press, 2014.

Nixon, Rob. *Slow Violence and the Environmentalism of the Poor*. Cambridge, MA: Harvard University Press, 2011.

Northcote, Rosalind. *The Book of Herbs*. London: John Lane/The Bodley Head, 1903.

"Nursing Echoes." *British Journal of Nursing* 61 (August 17, 1918): 113.

Nussey, Helen G. "The Hospital Almoner." In *The Fingerpost: A Guide to Professions for Educated Women*, 40–45. London: Central Employment Bureau for Women and Students' Careers Association, 1906.

———. *London Gardens of the Past*. With a preface by Eleanour Sinclair Rohde. London: John Lane/The Bodley Head, [1939] 1948.

Nussey, John T. M. "Walker and Nussey—Royal Apothecaries, 1784–1860." *Medical History* 14, no. 1 (1970): 81–89.

Opitz, Donald L. "'A Triumph of Brains Over Brute': Women and Science at the Horticultural College, Swanley." *ISIS* 104, no. 1 (March 2013): 30–62.

———. "Back to the Land: Lady Warwick and the Movement for Women's Collegiate Agricultural Education." *Agricultural Historical Review* 62, no. 1 (June 2014): 119–45.

Oppermann, Serpil, and Serenella Iovino. "Coda." In *Elemental Ecocriticism: Thinking with Earth, Air, Water, and Fire*, edited by Jeffrey Jerome Cohen and Lowell Duckert, 310–18. Minneapolis: University of Minnesota Press, 2015.

O'Sullivan, Paddy. "Struggle for the Vision Fair: Morris and Ecology." *Journal of the William Morris Society* 8, no. 4 (Spring 1990): 5–9.

Otto, Elizabeth, and Vanessa Rocco, eds. *The New Woman International: Representations in Photography and Film from the 1870s through the 1960s*. Ann Arbor: University of Michigan Press, 2011.

Oudit, Sharon. *Fighting Forces, Writing Women: Identity and Ideology in the First World War*. London: Routledge, 1994.

Ouida [Marie Louise de la Ramee]. "The New Woman." *North American Review* 158 (May 1894): 610–19.

———. *Moths*. Peterborough, ON: Broadview, 2005.

Palmer, Clare. "Stewardship: a Case Study in Environmental Ethics." In *The Earth Beneath: A Critical Guide to Green Theology*, edited by Ian Ball et al., 67–86. London: Society for Promoting Christian Knowledge, 1992. Reprinted in *Environmental Stewardship: Critical Perspectives, Past and Present*, edited by R. J. Berry, 63–75. London: T&T Clark, 2006.

Parham, John. "Was there a Victorian Ecology?" In *The Environmental Tradition in English Literature*, edited by John Parham, 156–71. Aldershot, UK: Ashgate, 2002.

Parkins, Wendy. *Mobility and Modernity in Women's Novels, 1850s–1930s*. Hampshire, UK: Palgrave Macmillan, 2009.

———. *Victorian Sustainability in Literature and Culture*. Edited by Wendy Parkins. London: Routledge, 2017.

Pemberton, Marilyn. *Out of the Shadows: The Life and Works of Mary De Morgan*. Cambridge: Cambridge Scholars, 2012.

———. "The Fairylands of Mary De Morgan: Seedbeds of Domestic Anarchy." *Latchkey: A Journal of New Woman Studies* 2, no. 2 (Winter 2010/11).
"Pharmacy and Medicines Bill." *British Medical Journal,* July 19, 1941.
Phillips, Dana. *The Truth of Ecology.* Oxford: Oxford University Press, 2003.
Plumwood, Val. *Feminism and the Mastery of Nature.* New York: Routledge, 1993.
Poovey, Mary. *Uneven Developments: The Ideological Work of Gender.* Chicago: University of Chicago Press, 1988.
Prendiville, Brendan. "British Environmentalism: A Party in Movement?" *Revue LISA/LISA* 12, no. 8 (2014): https://journals.openedition.org/lisa/7119.
"Presentation of French Garden in England to the King." *Times,* March 11, 1909.
Renzi, Kristen. "Dough Girls and Biscuit Boys: The Queer Potential of the Countercommunal Grotesque Body within Modernist Literature." *Modernism/Modernity* 22, no. 1 (2015): 57–80.
Richardson, Angelique. *Love and Eugenics in the Late Nineteenth Century: Rational Reproduction and the New Woman.* Oxford: Oxford University Press, 2003.
Richardson, Angelique, and Chris Willis, *The New Woman in Fiction and Fact: Fin de Siècle Feminisms.* Hampshire, UK: Palgrave Macmillan, 2002.
Rignall, John, and H. Gustav Klaus, eds. *Ecology and the Literature of the British Left: The Red and the Green.* London: Ashgate, 2012.
Rohde, Eleanour Sinclair. *A Garden of Herbs: Being a Practical Handbook to the Making of an Old English Herb Garden; Together with Numerous Receipts from Contemporary Authorities.* London: Medici Society, 1922.
———. *The Old English Herbals.* London: Longmans, Green, 1922.
Rossetti, Christina. *Christina Rossetti: Poems and Prose.* Oxford: Oxford University Press, 2008.
Rossetti, Dante Gabriel. *Collected Poetry and Prose.* Edited by Jan Marsh. Chicago: New Amsterdam Books, *1999.*
Rowbotham, Sheila. *Edward Carpenter: A Life of Liberty and Love.* London: Verso, 2008.
Rosowski, Susan J. "Willa Cather's Ecology of Place." *Western American Literature* 30, no. 1 (1995): 37–51.
Ruskin, John, E. T. Cook, and Alexander Wedderburn, eds. *The Works of John Ruskin.* London: George Allen, 1904.
Ryley, Peter. *Making Another World Possible: Anarchism, Anti-capitalism, and Ecology in Late Nineteenth- and Early Twentieth-Century Britain.* New York: Bloomsbury, 2103.
Sackville-West, Vita. *The Women's Land Army.* London: Imperial War Museum, 1997.
Sandilands, Catriona. "Lavender's Green? Some Thoughts on Queer(y)ing Environmental Politics." *Undercurrents* 6, no. 1 (1994): 20–24.
———. "Mother Earth, the Cyborg, and the Queer: Ecofeminism and (More) Questions of Identity." *National Women's Studies Association Journal* 9, no. 3 (1997): 18–40.
Sandilands, Catriona, and Bruce Erickson.eds. *Queer Ecologies: Sex, Nature, Politics, Desire.* Bloomington: Indiana University Press, 2010.
Sanecki, Kay N. *History of the English Herb Garden.* London: Ward Lock, 1992.
Schaffer, Talia. *The Forgotten Female Aesthetes: Literary Culture in Late Victorian England.* Charlottesville: University of Virginia Press, 2000.

———. "Nothing but Foolscap and Ink: Inventing the New Woman." In *The New Woman in Fiction and Fact: Fin de Siècle Feminisms*, edited by Angelique Richardson and Chris Willis, 39–52. Hampshire, UK: Palgrave Macmillan, 2002.

———. "The New Woman: Introduction." In *Literature and Culture at the Fin de Siècle*, edited by Talia Schaffer, 203–5. New York: Longman, 2007.

———. "The Victorian Novel and the New Woman." In *The Oxford Handbook to the Victorian Novel*, edited by Lisa Rodensky, 729–45. New York: Oxford University Press, 2013.

Schaffer, Talia, and Kathy Alexis Psomiades, eds. *Women and British Aestheticism*. Charlottesville: University Press of Virginia, 1999.

Scherer, Logan. "Entomology, Fiction, Intoxication: Annie Trumbull Slosson's Narratives of Obsession." *Legacy: A Journal of American Women Writers* 32, no. 2 (2015): 236–57.

Schreiner, Olive. *Woman and Labour*. New York: Frederick A. Stokes, 1911.

Seaton, Beverly. *The Language of Flowers: A History*. Charlottesville: University of Virginia Press, 1995.

Seymour, Nicole. *Strange Natures: Futurity, Empathy, and the Queer Ecological Imagination*. Champaign: University of Illinois Press, 2013.

Shiva, Vandana. *Water Wars: Privatization, Pollution, and Profit*, Cambridge, MA: South End, 2002.

———. *Who Really Feeds the World? The Failures of Agrobusiness and the Promise of Agroecology*. Berkeley: North Atlantic, 2016.

Showalter, Elaine. "Rivers and Sassoon: The Inscription of Male Gender Anxieties." In *Behind the Lines: Gender and the Two World Wars*, edited by Margaret Randolph Higonnet, Jane Jenson, Sonya Michel, and Margaret Collins Weitz, 61–69. New Haven: Yale University Press, 1987.

———. *Sexual Anarchy: Gender and Culture at the Fin de Siècle*. New York: Viking Adult, 1990.

Shteir, Ann B. *Cultivating Women, Cultivating Science: Flora's Daughters and Botany in England, 1760–1860*. Baltimore: Johns Hopkins University Press, 1999.

Singer, Charles. "Herbals." *Edinburgh Review* 237 (1923): 95–112.

Smith, C. "Macaulay, Dame (Emilie) Rose (1881–1958), author." *Oxford Dictionary of National Biography*, May 24, 2007.

Smith, Elise Lawton. *Evelyn Pickering De Morgan and the Allegorical Body*. New York: Fairleigh Dickinson, 2002.

Solnit, Rebecca. "Mysteries of Thoreau Unsolved: On the Dirtiness of Laundry and the Strength of Sisters." *Orion*, May/June 2013, 18–23.

Stein, Rachel. "The Place Promised that Has not Been: The Nature of Dislocation and Desire in Adrienne Rich's *Your Native Land/Your Life* and Minnie Bruce Pratt's *Crime Against Nature*." In *Queer Ecologies: Sex, Nature, Politics, Desire*, edited by Catriona Sandilands and Bruce Erickson, 285–308. Bloomington: Indiana University Press, 2010.

Stirling, A. M. W. *William De Morgan and His Wife*. New York: Henry Holt, 1922.

Stratigakos, Despina. "Female Firsts: Media Representations of Pioneering and Adventurous Women in the Early Twentieth Century." In *The New Woman International: Representations in Photography and Film from the 1870s through the 1960s*, edited by Elizabeth Otto and Vanessa Rocco, 56–70. Ann Arbor: University of Michigan Press, 2011.

Sutherland, Gillian. *In Search of the New Woman: Middle-Class Women and Work in Britain 1870–1914*. Cambridge: Cambridge University Press, 2015.

Syer, Geoffrey. "Morris and the Blunts." *Journal of the William Morris Society* 4, no. 4 (1981): 18–22.
Szerszynski, Bronislaw. "The Post-Ecological Condition: Irony as Symptom and Cure." *Environmental Politics* 16, no. 2 (2007): 337–55.
Talairich-Vielmas, Laurence. *Fairy Tales, Natural History, and Victorian Culture.* New York: Palgrave Macmillan, 2014.
Talbot, Meriel. "Editor's Letter to the Landswomen." *Landswoman,* August, 1918.
———. "Foreword." *Landswoman,* January 1918.
———. "A Message to the Land Army from Miss Talbot," *Landswoman,* May 1919.
Tarn, John Nelson. *Five Percent Philanthropy: An Account of Housing in Urban Areas Between 1840–1914.* New York: Cambridge University Press, 1973.
Taylor, G. R. S. "The Small Holdings Bill." *New Age* 1, no. 6. (June 6, 1907): 84–85.
Taylor, Jesse Oak. *The Sky of Our Manufacture: The London Fog in British Fiction from Dickens to Woolf.* Charlottesville: University of Virginia Press, 2016.
———. "Storm-Clouds on the Horizon: John Ruskin and the Emergence of Anthropogenic Climate Change." *19: Interdisciplinary Studies in the Long Nineteenth Century* 26 (2018): http://doi.org/10.16995/ntn.802.
———. "Where Is Victorian Ecocriticism?" *Victorian Literature and Culture* 43, no. 4 (2015): 877–94.
Thirsk, Joan. "Rohde, Eleanour Sophy Sinclair (1881–1950)." In *Oxford Dictionary of National Biography,* edited by Lawrence Goldman. Oxford: Oxford University Press, 2004.
———. *Alternative Agriculture: From the Black Death to the Present Day.* Oxford: Oxford University Press, 1997.
Thom, *Nice Girls and Rude Girls: Women Workers in World War I.* New York: I. B. Taurus, 2000.
Thorne, Isabel. "Food Production." *Times,* January 6, 1880.
Tilman, David, et al. "Global Food Demand and the Sustainable Intensification of Agriculture." *Proceedings of the National Academy of Sciences* 108, no. 50 (2011): 20260–64.
Todd, Barbara Euphan. "This Speaking Garden." *Spectator,* July 25, 1931.
"To the Members of the Women's Land Army." Address, July 1917, Bedfordshire Archives.
Tuana, Nancy. "Viscous Porosity: Witnessing Katrina." In *Material Feminisms,* edited by Stacy Alaimo and Susan Hekman, 188–214. Bloomington: Indiana University Press, 2008.
Turner, M. E., J. V. Beckett, and B. Afton. *Farm Production in England: 1700–1914.* New York: Oxford University Press, 2001.
Twinch, Carol. *Women on the Land: Their Story During Two World Wars.* London: Lutterworth, 1990.
Tylee, Claire M. *The Great War and Women's Consciousness: Images of Militarism and Womanhood in Women's Writings, 1914–64.* Iowa City: University of Iowa Press, 1990.
Uekoetter, Frank. *The Green and the Brown: A History of Conservation in Nazi Germany.* Cambridge: Cambridge University Press, 2006
Van Puymbroeck, Birgit. "Becoming a Land Girl: Reprinting Alice Meynell's 'The Shepherdess' in the *Landswoman.*" *Victorian Periodicals Review* 50, no. 2 (2017): 398–417.
Verdon, Nicola. "Business and Pleasure: Middle-Class Women's Work and the Professionalization of Farming in England, 1890–1939." *Journal of British Studies* 51, no. 1 (April 2012): 393–415.

———. "Left Out in the Cold: Village Women and Agricultural Labour in England and Wales during the First World War." *Twentieth-Century British History* 27, no. 1 (2016): 1–25.
Vicinus, Martha. *Independent Women: Work and Community for Single Women, 1850–1920.* Chicago: University of Chicago Press, 1985.
Vivan, Itala. "Geography, Literature, and the African Territory: Some Observations on the Western Map and the Representations of Territory in the South African Literary Imagination." *Research in African Literatures* 31, no. 2 (Summer 2000): 49–70.
Wagner, Lynne Shandi. "Seeds of Subversion in Mary De Morgan's 'The Seeds of Love.'" *Marvels and Tales* 29, no. 2 (2015): 245–64.
Wallace, Jo-Ann. "'How Wonderful to Die for What You Love': Mrs. Havelock Ellis's *Love-Acre* (1914) as Spiritual Autobiography." *Victorian Review* 37, no. 2 (Fall 2011):137–51.
———. "Edith Ellis, Sapphic Idealism, and *The Lover's Calendar* (1912)." In *Sapphic Modernities: Sexuality, Women and National Culture,* edited by L. Doan and J. Garrity, 183–200. New York: Palgrave Macmillan, 2006.
Warren, Karen. *Ecofeminism: Women, Culture, Nature.* Bloomington: Indiana University Press, 1997.
Warwick, Frances Evelyn Greville. *Coventry Herald,* July 21, 1899.
———. "The New Women and the Old Acres." *Women's Agricultural Times* 1 (1899): i.
———. *William Morris: His Homes and Haunts.* London: T. C. & E. C. Jack, 1912.
———. *A Woman and the War.* New York: George Doran, 1916.
Waters, Michael. *The Garden in Victorian Literature.* Aldershot, UK: Gower, 1988.
Waterton, Emma. *Politics, Policy, and the Discourse of Heritage in Britain.* New York: Palgrave Macmillan, 2010.
Watson, Jeanie. "'Men Sell Not Such in Any Town': Christina Rossetti's Goblin Fruit of Fairy Tale." *Children's Literature* 12, no. 1 (1984): 61–77.
Weltman, Sharon Aronofsky. *Ruskin's Mythic Queen.* Athens, OH: Ohio University Press, 1998.
White, Bonnie. *The Women's Land Army of the First World War Britain.* London: Palgrave Macmillan, 2014.
Whitman, Walt. *Leaves of Grass.* Philadelphia: Rees Welsh, 1882.
Wilkins, Mrs. Roland. "The Training and Employment of Educated Women in Horticulture and Agriculture, London: Board of Agriculture and Fisheries." *Journal of the Board of Agriculture,* 1915.
———. Letter to the *Times,* June 1, 1916.
Williams, A. Susan, Patrick Ivin, and Caroline Morse. *The Children of London: Attendance and Welfare at School 1870–1990.* London: Institute of Education, University of London, Bedford Way Papers, 2012.
Williams, Susan, and Tendayi Bloom. "Nussey, Helen Georgiana (1875–1965)." *Oxford Dictionary of National Biography.*
Williams, Raymond. *The City and the Country.* Oxford: Oxford University Press, 1973.
———. *Keywords.* New York: Oxford University Press, 1983.
Wilson, Cheryl. *Literature and Dance in Nineteenth-Century Britain: Jane Austen to the New Woman.* Cambridge: Cambridge University Press, 2009.
———. "Politicizing Dance in Late Victorian Women's Poetry." *Victorian Poetry* 46, no. 2 (2008): 191–205, 192.

Wintle, Sarah. "Horses, Bikes, and Automobiles: New Woman on the Move." In *The New Woman in Fiction and in Fact: Fin de Siècle Feminisms*, edited by Angelique Richardson and Chris Willis, 66–78. New York: Palgrave Macmillan, 2002.

Wisecup, Kelly. "Medicine, Communication, and Authority in Samson Occom's Herbal." *Early American Studies: An Interdisciplinary Journal* 10, no. 3 (2012): 540–65.

"Women and Agriculture." *Times*, 1899.

Wood, Gillen D'Arcy. "What Is Sustainability Studies?" *American Literary History*. 24, no. 1 (Spring 2012): 1–15.

Wood, Naomi. "Creating the Sensual Child: Paterian Aesthetics, Pederasty, and Oscar Wilde's Fairy Tales." *Marvels and Tales* 16, no. 2 (2002): 156–70.

Woolf, Virginia. *A Room of One's Own*. Edited by David Bradshaw and Stuart N. Clarke. Chichester, UK: Wiley Blackwell, 2015.

Yaeger, Patricia. "*Beasts of the Southern Wild* and Dirty Ecology." *Southern Spaces* 13 (February 2013): https://southernspaces.org/2013/beasts-southern-wild-and-dirty-ecology.

Youngkin, Molly. *Feminist Realism at the Fin de Siècle: The Influence of the Late Victorian Woman's Press on the Development of the Novel*. Columbus, OH: Ohio State University Press, 2007.

Zalasiewicz, Williams, et al. "Are We Now Living in the Anthropocene?" *GSA Today* 18, no. 2 (2008): 4–8.

Zweifel, Helen. "The Gendered Nature of Biodiversity Conservation." *National Women's Studies Association Journal* 9, no. 2 (1997): 107–23.

INDEX

aestheticism, 36, 65, 153, 155, 156–57, 159, 161, 184
agricultural depression, 1–2, 4, 8–9, 10, 15–17, 61, 66–68, 101, 111, 115, 123, 150, 165, 197n82
Alaimo, Stacy, 5, 44, 96, 132; *Bodily Natures,* 201n110; "Thinking as the Stuff of the World," 205n99; "Trans-Corporeal Feminisms," 80, 211n5
Albritton, Victoria and Fredrik, *Green Victorians,* 85, 202n3, 220n1
Allen, Grant, *The Typewriter Girl,* 13, 66
alternative agriculture, 2, 9, 15, 23, 60–83, 114, 124
Amherst, Alicia, 191, 192
anarcho-communism, 6, 14, 28, 71, 84, 87, 103
Andersen, Hans Christian, 57, 107
Andrews, Lucy, 101–2
Anthropocene, 3, 80, 83, 86, 99, 116, 117, 169, 187
anthropomorphism, 41, 54
Arber, Agnes, 4, 18, 24, 152–53, 155, 174, 181, 189; *Herbals: Their Origin and Evolution—A Chapter in the History of Botany, 1470–1670,* 161–65
Art Nouveau, 44
Arts and Crafts movement, 1–2, 14–18, 22, 27–59, 60, 62, 65, 71, 75, 77, 79, 82, 98, 134
Ashbee, C. R., 88–89
Atwood, Margaret, *The Handmaid's Tale,* 86, 117
aube de siècle, 22, 68, 83, 86, 134
Australia, 61
autonomy, 2–6, 28, 34, 38, 40–41, 49, 51, 53, 68–69, 72, 75, 80, 89, 106, 139, 155, 167, 173, 187

back to nature, 39, 134–35, 209n135, 213n64
back to the land, 2, 15, 23, 87, 101, 108, 116, 134, 212n15

Beaumont, Matthew, 156
bees, 13, 146, 154, 169–70
Beetham, Margaret, 136
Bennett, Jane, *Vibrant Matter,* 5, 10, 12, 21, 27, 41–42, 44, 54, 58, 76, 79, 154
Beowulf, 163
Bilston, Sara, 198–99n14
Bingham, Jane, 101
biodiversity, 5–6, 9–10, 25, 71, 82, 111, 113, 140, 151, 152, 154, 156, 164, 167, 175, 180, 182
Biodiversity Heritage Library, 19, 25, 187, 189, 193
Blatch, Harriot Stanton, 123
Blogg, G. D., 99
Blunt, Wilfred, 96
Bodley Head, 156–58
Bradley, Edith, 6
bread labor, 84, 87, 94, 104, 108, 136
Brexit, 151
Brighton Hove Museums, 190, 194
British Pharmacopoeia, 154
Bundy, Peggy and Bert, 104
Bunker, Daniel, 25
Burne-Jones, Edward, 27, 30, 49; *The Golden Stair,* 43; grandchildren, 27, 201n109
Burtt, John, 97–99, 103
Butler, Sharon, 104

Caird, Mona, 3; "A Defense of the 'Wild Woman,'" 12–13; "Marriage," 2; "Phases of Human Development," 10–12
Callen, Anthea, *Women Artists of the Arts and Crafts Movement,* 31, 58
Canada, 48, 61, 67
Carhullan, 86
Carpenter, Edward, 16, 23, 85, 88–89, 196n60; *Civilization,* 39; *Towards Democracy,* 96
Chamberlain, Hilda, 197n93
Chelsea Flower Show, 17, 163

INDEX

Chelsea Physic Garden, 17
Chemist & Druggist, 154, 177
Christie, Agatha, 4, 25, 152–54, 175; *Five Little Pigs*, 179–87, 189; "The Herb of Death," 176–79
Clarion, 15, 38
class: conservation and, 109, 209n135; eugenics and, 116, 119; leisure and, 7, 12; middle-class women and, 2–3, 5, 7, 9, 13, 16–17, 19, 62, 66, 69, 71, 74–75, 77, 85 105, 126, 211n5, 213n60; nature/culture divide and, 11, 61, 65, 96, 120; privilege, 5, 10, 18, 88, 16, 66; whiteness and, 126; women and, 2, 18, 124, 209; working class and, 7, 10, 126
Cobbett, William, 209n135
Cockerell, Olive J., 4, 13, 15, 18, 23, 27, 44, 189; and alternative agriculture, 60–83; and Octavia Hill, 1–2. *Illustrations:* "Bacteria Thrown Out of Work," 77; "Mending bell glasses," 74; *Oiketicus townsendi*, 20; "Planting seedlings," 72; "The Pool and the Tree," 45; "The Rain Maiden," 52; "We are two women," 64
Cockerell, Theodore, 83, 197n98
Cohen, Debra Rae, 132
Cohen, Jeffrey Jerome, *Elemental Ecocriticism*, 5, 37–38; "The Sea Above," 43
Collingham, Lizzie, 211n9
Collins, Tracy, 7
companion planting, 5, 61, 63, 69–78
conservation, 7, 18, 22–23, 25, 37, 50, 72, 89, 104, 108, 109, 200n48, 209n135
Conybeare, Dorothea, 149
Cornwall, 90, 93–94, 97
Cotswolds, 23, 84–85, 88–105
Crane, Walter, 27; "A Garland for May Day," 15
Cronon, William, "The Trouble with Wilderness," 210 n173
Croyden Brotherhood, 88, 99–100
Culpeper House, 155, 174
Cultivation of Lands Order, 120, 211n12
Cunningham, Gail, 7
Cynewulf, 163

Darre, R. Walter, 110
Darwin, Charles, 6, 21, 28, 77
Defence of the Realm Act, 120

Delvaux, Martin, 30
De Morgan, Evelyn Pickering, 22; *Daughters of the Air*, 38, 40; *The Storm Spirits*, 38, 40–41
De Morgan, Mary, 4–5, 12, 14, 22, 27–59; "Co-operation in England," 28; "The New Trades-Unionism and Socialism in England," 28; "The Pool and the Tree," 29, 42–49; "The Rain Maiden," 50–54; "The Seeds of Love," 29–36; "The Windfairies," 54–59
De Morgan, William, 27, 32, 35, 38–39
Denman, Lady Gertrude, 123
Dickens, Charles, 27
Dillon, Edward, 159
Duckert, Lowell, *Elemental Ecocriticism*, 5, 37, 38; "When It Rains," 53
Duncan, Isadora, 51, 57
Dyke, Thomas Hart, 163

ecological practices. *See* alternative agriculture; conservation
ecology, 6, 12, 18; dirty ecology, 148–49; ecological futurity, 152–88; elemental ecologies, 23, 27–60; natural history to science of, 19; pure ecology, 25; queer ecology, 5–6, 10, 13, 23, 44, 48, 64, 67, 71–72, 79, 84, 86, 90, 93, 96, 97, 107; utopian ecologies, 84–117
Egerton, George, 23, 84, 86, 89, 104; *Discords*, 107; *Keynotes*, 159; "The Regeneration of Two," 105–9
Eiloart, Arnold, 99–100
Eliot, George, 21
Eliot, T. S., 166
Ellis, Edith Havelock, 4, 6–7, 12, 13, 23, 84–97; *Attainment*, 89–93; "A Cornish Experiment in Cottages," 94–96; *The Lover's Calendar*, 96–97
Ellis, Havelock, 4, 23, 85, 196n60, 209n135
Emerson, Ralph Waldo, 88, 96
Entomologist, 19, 197n98
eugenics, 6, 18, 86, 92, 106, 108, 109, 113, 116, 134, 209n124, 209n137; in *Adnam's Orchard*, 109–16
Evesham, 61

Fabian Society, 88, 99
fairy tales, literary, 27, 29–59, 76–77

Fellowes, Anne, 190–91
Fellowship House, 87, 90, 93–95
Felski, Rita, 62
feminism, 5, 16, 86, 91–92, 115, 126, 132, 139, 212n51, 214n74
Flora (figure of), 15, 29–30, 41, 49, 68, 77, 79
floral discourse, 12, 31, 67, 112, 114, 115, 165
floral ecologies, 1, 159, 165
flower gardening, 9, 27, 74, 111, 113, 159, 166
foraging, 99, 153, 164, 183
Fowler, James, 199n4
food: environmental damage of contemporary foodways, 26, 60–83; food security and link to war, 16; importation and outsourcing of, 1, 6, 13; shortage of local food, 1–2, 14; and urban gardens, 193; women and local growing of, 1–2, 7, 10, 15–16, 23; and Women's Land Army, 118–51
food justice, 5, 13, 15, 17, 18
French gardening, 60–83

Gaard, Greta, 48
Gagnier, Regenia, 3–4
Garden City, 15, 62
Gaskell, Elizabeth, 158, 209n135
Gates, Barbara T., *Kindred Nature*, 26, 204n83, 205n106
George IV, 190, 220n5
Gerard, *Herball*, 2, 171, 195n11
Gifford, Terry, 214n92
gifts, 193
Gilead, 86
Girton College, 16, 71
Glover, Mary, 99
Goethe, Johann Wolfgang von, 87, 95
Goodwill Dressmakers of the Croyden Brotherhood, 100
Gould, Peter, *Early Green Politics*, 213n64
Grand, Sarah, 3–4, 7, 23, 84, 189, 209n135; *Adnam's Orchard*, 109–17, 122, 124, 133–34; *The Heavenly Twins*, 79; "The New Aspect of the Woman Question," 12
Grayzel, Susan, 211n8
Great War, 1–2, 4, 8–10, 14, 17, 22–24, 61, 66, 74, 82, 101; and the Women's Land Army, 118–51
Greenham Common, 86, 193

Greenway, Judy, 86, 94, 104, 207n65
Grieve, Maud, *A Modern Herbal*, 4, 9, 14, 17–18, 24, 152–87; the Whins Medicinal and Commercial Herb School and Farm, 152, 163

Hager, Lisa, 108
Hall, Sarah, *Daughters of the North*, 26, 86, 117
hands, 1, 8, 16, 21, 29, 34–35, 44, 66, 72–77, 79, 80–82, 84, 105, 131
Hardy, Dennis, 87
Hardy, Thomas, *Tess of the d'Urbervilles*, 30
Heilmann, Ann, *New Woman Hybrities*, 79
Helmreich, Anne, 111
Hekman, Susan, 212n51
Henderson, Frank and James, 99
herbalism, 8, 24, 152–87
herbal revival, 152–87
herbs: English lavender, 171–73; foxglove, 153, 154, 169, 170; henbane, 24, 153, 171; lavender, 24; Seal lavender, 13, 155
Hewer, Dorothy, 173
Hill, Octavia, 1, 7, 10, 16, 22, 23, 59, 60, 73–74, 80, 140
Hinton, James, 23, 85
Hockin, Olive, 2, 4, 5, 7, 13, 15, 24, 66, 118, 119, 121–22, 129; "A Cobwebbed Woodland," 134; *Two Girls on the Land: War Time on a Dartmoor Farm*, 133–39
Holmes, E. M., 154–55
homophobia, and discursive formations of nature, 10
Hoyles, Martin, 82

Independent Labor Party, 88
intensive culture, 1, 4, 7–9, 12, 14, 23, 60–83, 109–16
International Congress of Women, 16

Kant, Immanuel, 87, 92, 95
Kenworthy, John C., 99
Kew Gardens, 17, 47
King, Amy, 12, 29; *Bloom*, 30, 70, 74, 159
King, Peter, *Women Rule the Plot*, 197n93
Kirkpatrick, Lily, 85, 90, 94, 96, 97
Knechtel, Ruth, "Olive Schreiner's Pagan Animism," 198n108
Kortsch, Christine Bayle, 98

Kropotkin, Prince Peter, 14, 23, 28, 66, 85, 124; *The Conquest of Bread*, 16, 71; *Mutual Aid*, 29, 195n31

Laith, Randy, 217n61
Land Question, 2, 15, 18, 119, 123–28, 190
Landswoman, 4, 24, 119, 122–23, 127, 128–39
Lane, John, 156
Lawrence, D. H., "The Fox," 67
Ledger, Sally, 3
Leighton, Fredric, 30, 38
Leopold, Aldo, 50
Leyel, Hilda, 17, 24, 152, 155, 163, 174–75, 180
Lobb, Mary, 13
Lullingstone Castle, herb garden, 163–64

Macaulay, Rose, 4–5, 9, 18, 24, 119–22; "On the Land," 140–50
MacCarthy, Fiona, *The Simple Life*, 202n3
Marsh, Jan, 29–30, 58; *Back to the Land*, 213n63; "Concerning Love: News from Nowhere and Gender," 198n11; *Jane and May Morris: A Biographical Story*, 199n27
Marx, Karl, 11, 30, 91
Mayer, Edward, 206n14
McKay, C. D., *The French Garden: A Diary and Manual of Intensive Culture*, 61–62
McKibben, Bill, 89
Merchant, Carolyn, 217n70
Merry England, 127–28, 185
Mew, Charlotte, 62
Meynell, Alice, 133
Miller, Elizabeth: *Framed*, 35, 199n40; *Slow Print*, 202n3
mobility, 3, 6, 17, 21–22, 28–30, 40–41, 49, 56, 59, 62, 100, 127, 146, 153, 168
Morris, Jane, 29, 31, 62, 81, 83, 199n27
Morris, May, 4, 13, 22, 58, 199, "Honeysuckle," 38–40
Morris, William, 23, 28, 30–31, 37, 39, 66, 71, 85, 91, 95–96, 110, 123, 209n135; "The Art of the People," 37; "The Dawn of a New Epoch," 31; and Gerard's *Herball*, 195n11; *News from Nowhere*, 14–15, 31, 198n11
Mortimer-Sandilands, Catriona, 46

Morton, Timothy, *The Poetics of Spice*, 10
moths, and the New Woman, 21–22
Muñoz, José Esteban, 23, 107; *Cruising Utopia*, 85; "ecstatic time," 85, 97; "straight time," 6, 85–86, 93, 107
Murphy, Patricia, 30

Naeem, Shahid, 25
National Trust, 104, 209n137
natural history, 18, 46–48, 164
nature writing, 25–26, 83, 142
neo-paganism, 133, 135, 165
New Age, 15
New Life, 28, 84–117; and Edward Carpenter, 196n60
Newnham College, 4, 118, 161, 162
New Order, 15, 99, 100
New Mexico, 19, 21
New Woman: in alternative agriculture, 2, 9, 15, 23, 60–83, 114, 124; and autonomy, 2–6, 34, 38, 40–41, 49, 51, 53, 68–69, 72, 75, 80, 89, 136, 139, 155, 167, 173, 187; criminality of, 34–36, 153–54, 175–76, 179, 181–83, 185, 199n40; in Great War, 118–51; and image in print culture and photography, 7–8, 24, 27, 40, 66–67, 98–100, 124, 126, 129, 132, 149; and rational dress/clothing, 3, 8, 69, 85, 87, 95, 97–105, 127, 129, 159
New Zealand, 61
Nietzsche, Friedrich, 105
Nixon, Rob, *Slow Violence*, 47, 115, 209n136
Northcote, Lady Rosalind, 24, 153, 155–62, 164, 169, 171, 174, 189, 191, 216n26, 220n5
Nussey, Helen: bequeathal of herbstrewer's dress, 190–94; *A French Garden in England*, 1–2, 4–5, 8, 12–15, 21, 23, 60–83, 121, 124, 135, 189; *London Gardens of the Past*, 192–93

Ouditt, Sharon, 213n60

Packard, A. S., *The Bombycine Moths of North America*, 197n98
Palmer, Clare, 211n11
Pankhurst, Adela, 15, 197n81
Pankhurst, Emmeline, 210n2
Pankhurst family, 118

Paris, 61
Parkins, Wendy, 22
Pemberton, Marilyn, xi, 199n27
Peter Rabbit, 77
Pharmacy and Medicines Act of 1941, 25, 154, 174–76, 179, 183, 186
Phelps, Elizabeth Stuart, "Zerviah Hope," 109
plant collection, 28, 43, 47–48
plant literacy, 2, 152–53, 156, 157, 162, 164, 167, 170, 175–76, 182–83, 187
Plumwood, Val, 11, 132, 212n51
Poisons Act of 1933, 25, 179, 182
Pomona, 30, 41, 49
Prendiville, Brendan, 209n120
Punch, 7–8, 27, 66–67, 124, 126

queer, tragic stereotype of, and the New Woman, 67
queer ecology, 5–6, 13, 23, 44, 48, 64, 68, 71–72, 79, 96–97
queer utopia and future, 64, 85–86, 90, 93, 107, 117

rational dress, 85, 87, 98, 100, 104, 97–105
Rayne, Bertha, 150
Redistribution Bill of 1885, 112
Richardson, Angelique, 209n135
Rohde, Eleanour Sinclair, 4, 24, 82, 152, 155, 161–66; *A Garden of Herbs*, 161, 165; *The Old English Gardening Books*, 161; *The Old English Herbals*, 161, 163; *The Scented Garden*, 161
Rossetti, Christina: "A Birthday," 97; "Goblin Market," 77
Rossetti, Dante Gabriel, *The Day Dream*, 29
Rothschild List, 104, 109, 209n133
Ruskin, John, 9, 83; and ideal of manual labor, 14, 73, 202n3; "The King of the Golden River," 38; "On the Nature of the Gothic," 110; "Of Queen's Gardens," 32, 112–13, 155, 165, 198n14; and simplification, 85; "Storm Clouds of the Nineteenth Century," 37

Sackville-West, Vita, 122, 163
Sanecki, Kay, 168
Schaffer, Talia, 3, 114, 159, 161

Schreiner, Olive, 3, 13, 118
Secret Garden, 165
sexual anarchy, 3, 41, 74, 155, 176, 182, 184, 186
Shakespeare, William, 158; *Hamlet*, 171; "Sonnet 5," 160, 185
Shaw, Nellie, 4, 6, 23, 84–85, 87–88, 94, 97–105
Shiva, Vandana, 9, 211n10
Showalter, Elaine, 3
Shteir, Ann, 201n106
"Simple Life" and simplification, 9, 23, 62, 65, 78, 87–90, 92, 94–95; and the New Life, 84–117
Sinclair, William, 102
Smith, Elise Lawton, 40–41
Smith, Naomi, 141
Smith, W. H., & Sons, 15, 62
socialism, 28, 31, 85, 95, 113, 114; early green, 1, 2, 4, 6, 9, 14–18, 22, 27–59, 62, 71, 79, 110, 118, 123–24, 133–35, 139, 155, 190
soil, 1, 5, 7–9, 12, 16, 21, 24, 34, 63–65, 67, 69, 71–73, 75–79, 101, 103, 115, 119, 122, 124, 126, 129, 130, 135–36, 144, 148–49, 151, 153, 166, 168, 172, 211n10
Spencer, Herbert, 11, 85
St. John, Arthur, 99
Stead, W. T., 15, 62
Studley College, 4, 14, 16, 71, 108, 118, 123, 150
Sutherland, Gillian, 9
Swanley College, 16–17, 127, 150
Szersynski, Bronislaw, 63

Talbot, Meriel, 118, 127–28, 133–38, 148, 150, 212n15
Taylor, Jesse Oak: *The Sky of Our Manufacture*, 219n1; "Storm Clouds," 37
Taylor, Una Ashworth, "The Seed of the Sun," 156–57
Thackeray, William Makepeace, *The Ring and the Rose*, 63
Third Reform Act, 112
Thirsk, Joan, 15; *Alternative Agriculture*, 203n11, 205n108
Thomas, William Beach, 61
Thoreau, Cynthia, 102
Thoreau, Henry David, *Walden*, 23, 85–86, 88, 96, 102

Tolstoy, Leo, 23, 85, 87–88, 98–99, 101, 103, 135
Tuana, Nancy, 5
Turner, M. E., 210n148

Uekoetter, Frank, 210n173
United States, 61, 88, 123, 209n135
Utopia, 2, 4, 6, 14, 23, 26, 28–30, 36, 39–40, 62, 64, 68–69, 76, 84–118, 154, 166, 169, 194, 198n11

vegetarianism, 65, 71, 85
Verdon, Nicola, 6
Vicinus, Martha, 9

Wallace, Jo-Ann, 96–97
water, 1, 14, 26, 28–29, 36–59, 102–3
Walker, Sarah Ann, 190
Warner, Sylvia Townsend, *Lolly Willowes*, 68
Warwick, Lady Frances Evelyn, 4, 6, 8, 14, 16–18, 22, 60, 62, 71, 108, 118, 122, 123–24, 126–27
Waterton, Emma, 209n137

Wells, H. G., 97–98
Weltman, Sharon Aronofsky, 198n14
Whins Medicinal Herb School and Farm, 152, 163
White, Bonnie, 212n17
Whiteway: A Colony on the Cotswolds, 97–105
Whitman, Walt, 23, 85, 92–93, 97, 95–96
Wilde, Oscar, 27, 85–86, 97, 156
Wilkins, Louisa, *Agricultural Education for Women*, 118, 127
Wilkinson, Frances, 17, 127
Woman Question, 2, 12, 18, 112, 124
Women's Agricultural and Horticultural International Union (WAHI), 17
Women's Agricultural Times, 8
Women's Land Army, 2, 4, 8, 17, 23–24, 117, 118–51
Woolf, Virginia, 66
Wordsworth, William, 209n135

Yaeger, Patricia, 148–49
Yellow Book, 156, 159
Yonge, Charlotte, 158

RECENT BOOKS IN THE SERIES
Under the Sign of Nature: Explorations in Ecocriticism

Dan Brayton
Shakespeare's Ocean: An Ecocritical Exploration

Jennifer K. Ladino
Reclaiming Nostalgia: Longing for Nature in American Literature

Byron Caminero-Santangelo
Different Shades of Green: African Literature, Environmental Justice, and Political Ecology

Kate Rigby
Dancing with Disaster: Environmental Histories, Narratives, and Ethics for Perilous Times

Adam Trexler
Anthropocene Fictions: The Novel in a Time of Climate Change

Eric Gidal
Ossianic Unconformities: Bardic Poetry in the Industrial Age

Jesse Oak Taylor
The Sky of Our Manufacture: The London Fog in British Fiction from Dickens to Woolf

Michael P. Branch and Clinton Mohs, editors
"The Best Read Naturalist": Nature Writings of Ralph Waldo Emerson

Lynn Keller
Recomposing Ecopoetics: North American Poetry of the Self-Conscious Anthropocene

Julia E. Daniel
Building Natures: Modern American Poetry, Landscape Architecture, and City Planning

Serenella Iovino, Enrico Cesaretti, and Elena Past, editors
Italy and the Environmental Humanities: Landscapes, Natures, Ecologies

Christopher Abram
Evergreen Ash: Ecology and Catastrophe in Old Norse Myth and Literature

Elizabeth Hope Chang
Novel Cultivations: Plants in British Literature of the Global Nineteenth Century

Emily McGiffin
Of Land, Bones, and Money: Toward a South African Ecopoetics

Alicia Carroll
New Woman Ecologies: From Arts and Crafts to the Great War and Beyond

www.ingramcontent.com/pod-product-compliance
Lightning Source LLC
Chambersburg PA
CBHW030824230426
43667CB00008B/1366